SOVIETS IN SPACE

KOSMOS
A series exploring our expanding knowledge of the cosmos through science and technology and investigating historical, contemporary and future developments as well as providing guidance for all those interested in astronomy.

Series Editor: Peter Morris

Already published:

Asteroids Clifford J. Cunningham
The Greatest Adventure Colin Burgess
Jupiter William Sheehan and Thomas Hockey
Mars Stephen James O'Meara
Mercury William Sheehan
The Moon Bill Leatherbarrow
Saturn William Sheehan
The Sun Leon Golub and Jay M. Pasachoff
Uranus and Neptune Carolyn Kennett
Venus William Sheehan and Sanjay Shridhar Limaye
Soviets in Space Colin Burgess

SOVIETS
IN
SPACE

*Russia's Cosmonauts
and the Space Frontier*

COLIN BURGESS

REAKTION BOOKS

To the lost crewmembers of the Soyuz programme,
who gave their lives while returning home from space

Published by Reaktion Books Ltd
Unit 32, Waterside
44–48 Wharf Road
London N1 7UX, UK
www.reaktionbooks.co.uk

First published 2022
Copyright © Colin Burgess 2022

Printed and bound in Great Britain
by TJ Books Ltd, Padstow, Cornwall

A catalogue record for this book is available from the British Library

ISBN 978 1 78914 632 5

CONTENTS

PROLOGUE

Throughout history, our shared and fanciful dreams of flying to the Moon and the known planets have both intrigued and tormented humankind. Over more recent centuries our knowledge of the sheer magnitude of the universe and Earth's position in it has rapidly expanded, along with the ongoing development of science and technology. It was hoped that these factors would one day combine to lead us towards taking those first baby steps out into the wonders – and the known and unknown terrors – of space.

Flights beyond our atmosphere and the seemingly insurmountable grip of gravity would remain the subject of conjecture, myth and science fiction until it was finally recognized that the only viable means of defeating the forces of gravity rested with propulsion devices capable of carrying vehicles into space. The word most appropriate to this was 'rockets'. Once some practical means of conveying passengers on rocket flights beyond our atmosphere had been established, life-supporting arrangements such as providing breathable air and protective garments would follow.

Year after year, studies and advances in space science and astronautics continued, turning the yellowing pages of fiction into the realms of achievable fact. Unlike engines of the past, which had been reliant upon the wheel and a wheel-like screw for locomotion on land, water and in the air, it would be rockets relying on Isaac Newton's third law that would finally allow us to overcome the frustrating problem of gravity. This law stated that for every action (force) in nature, there is an equal and opposite reaction. By observing this immutable law of physics, unmanned satellites, planetary probes and even crewed spacecraft could one day be launched atop powerful rockets capable of tearing a steady path through Earth's surprisingly thick atmosphere, venturing out into the largely inhospitable grandeur of space.

The invention of gunpowder and its use in skyrockets is rightfully attributed to the Chinese, dating back more than 2,000 years. They found that a combination of charcoal, saltpetre and sulphur created an explosive propellant then known as 'black powder'. This resulted in considerable international interest in converting these novelty rockets into an instrument of war, rather than exploration. Not surprisingly, even the Chinese employed rudimentary rockets for military purposes, such as when the town of Kaifeng was besieged by Mongol invaders in 1232. The fundamentals of rocketry are even believed to have reached as far as Italy and France sometime around the thirteenth century. Russian history records the use of rockets in 1516 near the Russian (now Ukrainian) city of Belgorod-Kievsky, following repeated incursions by Crimean Tatars, while other rockets are said to have been sighted in the Russian city of Nikolskoye (now Ussuriysk) around 1675.

Following the development and use of military missiles in Europe, the 'Rocket Enterprise' (*Raketnoe Zavedenie*) was reportedly established in Moscow around 1680. According to their archival records, a signalling rocket developed in the St Petersburg Arsenal in 1717 was capable of reaching an altitude of several hundred metres. In 1732 the arsenal produced twenty rocket-launching devices for the Russian border fortress of Brest.

There were many pioneers in the early field of Russian rocketry and its possible application to space travel. One of the more prominent advocates was military artillery engineer General Konstantin Ivanovich Konstantinov, born in St Petersburg in 1818. Recognized as a founder of experimental rocketry, he had studied the science of ballistics and was convinced that rockets might one day be used as vehicles devoted to travelling into space. In 1847 Konstantinov created a ballistic rocket pendulum, which would, in time, establish a law of changing rocket motion. With the help of this device, he was able to determine the form and design of a rocket based on its ballistic characteristics. In 1849 Konstantinov took over as commander of the St Petersburg 'Rocket Enterprise' Arsenal and subsequently spent several years in Europe studying rocket development. In 1861 he led the foundation of the Nikolayev rocket production plant, which became partially operational in 1867. The plant continued to grow, but Konstantinov would

Gustave Doré, 'Baron Munchausen's Voyage to the Moon', 1868,
engraving depicting a novel means of flying to the Moon.

General Konstantin
Konstantinov, 1858.

never live to see it completed. On 12 January 1871 he died suddenly at
the age of 52.[1]

Earlier, in 1804, a British Army officer named William Congreve
had designed a stick-guided rocket that could carry either an incendiary
device or anti-personnel warhead into enemy territory up to 3 kilometres (2 mi.) distant. These were first utilized two years later and fired
as weapons in conflicts across Europe and the United States. During
the war of 1812, British forces used Congreve rockets against American
defenders at the Battle of Fort McHenry. It was this assault that inspired
Francis Scott Key to write his poem 'The Defence of Fort M'Henry',
which was later set to music as 'The Star-Spangled Banner' and adopted
as the national anthem of the United States. Mention is made within
the anthem of 'the rocket's red glare'. By 1815 the British Army had
formed its own rocket brigade and, that year, rockets would play a
major role during the Battle of Waterloo in Belgium.

Among all the Russian scientists involved in that nation's history
of early rocketry, one name stands tall: Konstantin Eduardovich
Tsiolkovsky. A self-educated man, deaf since the age of nine, this
creative and prodigious genius was intrigued by the prospect of space
travel. Tsiolkovsky not only explored the basic principles of rocket
propulsion and the properties of the atmosphere, but designed multistage rockets. By 1929 he had calculated that for a single-stage rocket

to achieve what he described as a 'cosmic velocity', the weight of the fuel had to exceed that of the rocket by at least four times. He therefore set about designing a remarkable vehicle he called his 'rocket train'.[2] This comprised a series of rocket stages mounted on top of each other. During the launch of this vehicle, each stage would ignite and burn until its fuel was exhausted, at which time it would be jettisoned and the next segment fired to continue the upward thrust. The last remaining stage would carry the payload to its final destination. It was the same principle later used in such mighty multistage rockets as Saturn v, which carried America's Apollo astronauts to the Moon.

The German physicist and engineer Hermann Oberth (1894–1989) was another visionary considered a doyen and founding father of rocketry and astronautics, and he had been drawn to the same conclusions as Tsiolkovsky. 'If there is a small rocket on top of a big one,' he wrote, 'and if the big one is jettisoned and the small one is ignited, then their speeds are added.'[3]

Konstantin
Tsiolkovsky,
photographed *c.* 1930.

Tsiolkovsky also sketched out concepts for Earth-orbiting satellites, as well as space platforms that could become an orbiting waystation for flights to the Moon and near planets. Furthermore, he worked on ways of returning space vehicles safely through the blistering heat of re-entry and the use of solar energy in powering artificial satellites. He even constructed a wind tunnel, enabling him to study the aerodynamic characteristics of differently shaped objects. Tsiolkovsky also theorized that rockets could be fuelled by combining liquid oxygen with such substances as alcohol, liquid hydrogen, hydrocarbons, methane and kerosene. And yet, this amazing man of science and rocketry received very little recognition for his efforts from his own country and was unknown outside of Russia. 'How difficult it is,' he once wrote, 'to work for years all alone under unfavourable circumstances and not see any light or help from anywhere.'[4]

Nevertheless, Tsiolkovsky patiently continued his work, which slowly began to garner attention and acceptance. By 1919 he had gained sufficient recognition to be elected to the Socialist Academy, the goals of which were research in the social sciences, history and the theory and practice of socialism. Between 1925 and 1932 he produced some sixty works dealing with such diverse subjects as astronautics, astronomy, mechanics, physics and even philosophy. He died two days after his 78th birthday. These days, he is considered to be one of the leading founding fathers of modern rocketry and astronautics.

Apart from Tsiolkovsky and Oberth, there were many other celebrated names from several countries who played a key role in the rapidly developing history of rocketry and astronautics. Among these were the Americans Robert Goddard and Frank Malina, the Frenchman Robert Esnault-Pelterie, the German American Willy Ley and the Germans Konrad Dannenberg, Walter Dornberger and Wernher von Braun.

During the Second World War, all of the combatants used rockets in many different forms as weapons, but, driven by desperation, it was Germany that made gigantic steps in the development of rocket-powered guided missiles. These were the country's *Vergeltungswaffen*, or Vengeance Weapons, and the first to be developed was the jet-propelled v-1, the forerunner of modern cruise missiles. Limited in range, they were launched towards England in their thousands, beginning on 13 June 1944, from sites situated along the French and Dutch coasts. Besieged Londoners came to know these frightening and destructive missiles as buzz bombs or doodlebugs.

German V-2
mounted on a
mobile launch pad.

There was even worse to come, as Germany's rocket scientists and engineers developed the much larger v-2 missile at the Peenemünde Army Research Centre, located by the Baltic Sea. This work, involving a slave labour force of around 12,000, was headed by rocket experts Walter Dornberger and Wernher von Braun. The liquid-propelled v-2 was first launched against Allied cities on 6 September 1944, with targets including London, Antwerp and Liege experiencing a reign of terror during which around 9,000 civilians and service personnel were killed and city areas were reduced to rubble.

As an Allied victory became increasingly certain, and troops moved relentlessly through a collapsing Germany, teams were formed from the United States, the Soviet Union and the United Kingdom tasked with locating and recovering as many intact and partial rockets as possible, while also taking into custody the scientists and technicians responsible for the development of these highly destructive weapons. Each country wanted to develop or increase their own military missile programmes, with the seized rocketry and personnel subsequently shipped overseas to missile operation centres. The United States alone, through Operation Paperclip, was able to capture sufficient v-2s and associated hardware

An American soldier examines the upper end of a v-2 captured on a train.

to construct approximately eighty rockets back on American soil. It also welcomed the voluntary surrender to the u.s. forces by Wernher von Braun and his rocket team, along with information he later revealed on where they had concealed their blueprints pertaining to the v-2 so these would not fall into Soviet hands. Meanwhile, the Soviet search teams also located and seized possession of v-2 manufacturing facilities, later using these to study the construction of the rockets, and began work on modifying the missiles back in the Soviet Union. By October 1948 they had constructed their own more powerful derivative of the v-2, which they called the R-1, and had successfully launched the first of these at their rocket development and launch site in Kapustin Yar, east of Volgograd (formerly Stalingrad).[5]

With the end of the Second World War, rockets and rocket planes now began to reach and exceed new milestones, climbing ever higher and faster into the atmosphere. All too soon, insects, plants, spores, seeds and some small creatures including mice would be launched atop modified v-2 missiles, allowing scientists to investigate the effects of massive acceleration forces of launch and re-entry, as well as brief periods of weightlessness, on various life forms. Very little was known of these effects, and the data produced was subjected to intense study before the first human subjects could undertake those same risks.

Soon, both nations would be in a position to take those first tentative steps out into our new frontier, and the dramatic, competitive era that came to be universally known as the 'Space Race' would begin.

Many decades ago, while postulating on his nation's future in space exploration, a contemplative space theorist, Konstantin Tsiolkovsky, wrote, 'I have no difficulty imagining the first man overcoming the Earth's gravity and rushing into space. He is a Russian, a citizen of the Soviet Union; his trade, most probably, is a pilot. He is courageous, yet void of any recklessness. I see his frank Russian face.'[6]

At the time he wrote those words, a small Russian child named Yuri Gagarin was about a year old. Twenty-five years after Tsiolkovsky's death, his prophetic thoughts would be transformed into fact.

1

PUPNIKS AND SPUTNIKS

Many of the earliest Soviet space explorers were born into a life of seclusion and deprivation on the back streets of Moscow, unloved and passing each wretched day scrounging food and finding shelter where they could. They had become inured to living rough and tough under often appalling weather conditions, and yet they would one day become known and lauded throughout the Soviet Union and even around the world. Most of them were destined to find themselves strapped inside a rocket heading into the upper atmosphere, and some into Earth's orbit. They may have been tiny, mixed-breed dogs, but their names and faces would soon grace newspaper and magazine covers. Sadly enough, a few would also perish as two great superpowers engaged in a massive effort to achieve international prestige by sending the first human beings into the largely unknown, endless arena of space.

SOMETIME IN THE WINTER of 1949, the telephone rang in the office of Vladimir Yazdovsky, a professor at Moscow's Institute of Aviation Medicine (IAM). At the time, he was busily engaged in biomedical studies, and when he answered the call he did not recognize the name of the person introducing himself at the other end of the line, Sergei Korolev. Then again, Korolev's name was known only to a privileged few, as he was the lead engineer heading the Soviet Union's thrust into space. It was not until after his untimely death in 1967 that his identity was finally revealed as the mysterious Chief Designer of the Soviet space programme.[1]

In 1945, 38-year-old Sergei Pavlovich Korolev, an aircraft and rocket designer and then a colonel in the Russian Army, was just one of several engineers and technicians sent into post-war Germany to study and test v-2 rockets and work on their later variants. Months later, with precious little notice, they were ordered back to the Soviet Union to work over the next seven years with hundreds of captured German rocket technicians in developing the nation's own missile programme. By early 1947 everything and everybody was in place, and a frenzied period of rocket activity had begun. The first launch of a v-2 rocket (now renamed the R-1) from Soviet soil took place on 30 October that year, lifting off from a secret launch base in Kapustin Yar, located some 100 kilometres (62 mi.) east of Volgograd. It would fly 300 kilometres (186 mi.) downrange before arcing down and slamming back to the ground in the designated target area. It was the best possible start.

When Korolev contacted Yazdovsky at the IAM, the Chief Designer had been involved in sending small biological specimens into the upper atmosphere on R-1 rocket flights. Now Korolev wanted to up the

Chief Designer Sergei Korolev with one of the space dogs.

tempo, as the Americans had already begun sending small monkeys aloft in their own captured v-2 rockets. These animals were encased in rudimentary canisters secured within the nose cones of the former missiles of brutal destruction. The prospect of working with Korolev on upgraded biological flights into space proved fascinating to Yazdovsky, and they would soon set out certain objectives, including the type of animal subjects to be used.[2]

Taking into account the earlier work with canines of renowned scientist and physiologist Ivan Pavlov, they decided that dogs could be easily trained and were far more placid than unpredictable and fidgety monkeys. They also had to be small enough (around 6–7 kg or 13–15 lb) to fit into a special hermetically sealed capsule that could be inserted into the nose cone of an R-1. Another factor was the sex of the canine subjects and the problem of waste matter. A urine and faeces collection tube could be easily fitted to a female's rear end, as they normally squatted to urinate, whereas male dogs would customarily need to lift one of their legs, which was impractical in the confined space of the capsule. Armed with this information, Yazdovsky sent out some of his biomedical team to scour the streets of Moscow for likely candidates

Sergei Korolev (centre) with Vladimir Yazdovsky (in uniform). On the left is Nikolai Kamanin. From 1960 to 1971, General Kamanin would be the head of cosmonaut training for the Soviet space programme.

– which, fortunately for him, were quite plentiful – knowing that these dogs would be street-hardened, and therefore more likely to survive the rigours and stresses associated with a rocket flight than a domesticated animal.

As their training began, several dogs would prove to be uncooperative and were removed, while others began to fall into two categories. Some were found to be better suited to the shorter ballistic missions, while others, quieter and more patient by nature, would be groomed for later orbital missions.

The training was thorough; it involved the animals standing still for lengthy periods, adapting over time to wearing restricting, specially made canine spacesuits and being confined within small canisters for up to three weeks at a time. They would also be strapped into centrifuges and whirled around to simulate the explosive acceleration, noise and vibrations associated with a rocket launch. As on possible future flights, all of their reactions and vital signs were carefully monitored through sensors attached to their bodies. Once Yazdovsky felt that the first candidates were ready, he informed Korolev, and the first launch with canine passengers could proceed.

In the predawn light of 22 July 1951, an R-1B rocket stood at the ready at the Kapustin Yar launch site. Earlier, Yazdovsky had assisted in loading two small dogs named Tsygan (Gypsy) and Dezik (simply a pet name) into their sealed capsule, which was situated in the tapered upper section of the R-1B. He patted the two placid dogs before sealing the hatch.

The lift-off went as planned, with the rocket reaching an altitude of around 110 kilometres (68 mi.), following which the nose cone separated and fell back towards the ground. The parachute deployed and billowed out with a snap, slowing the nose cone, which touched down without incident. The fifteen-minute flight was deemed successful, and once the recovery team had reached the landing site, the two dogs emerged from their capsule understandably excited but unharmed.[3] Aleksandr Seryapin, who opened the hatch to release Tsygan and Dezik, was delighted. 'The first flight turned out very successful,' he recalled. 'The dogs were alive. When we released them, a lot of cars pulled up, and Sergei Pavlovich Korolev was in one of them. When he saw the dogs – in my opinion, there wasn't a happier person there. He grabbed them, ran around the cabin with them, poured them water [and] gave them sausages and sugar.'[4]

A Soviet technician with Tsygan (left) and Dezik, who made the first suborbital Soviet space flight on 22 July 1951.

Following the success of the first flight, Dezik would be sent up again a week later, on 29 July, this time with a canine companion named Lisa. While essentially similar to the earlier rocket flight, on this occasion the parachute malfunctioned and both dogs were killed on impact with the ground. Tsygan would fare much better. Following her only ride aboard a rocket, she was adopted as a pet by the Soviet physicist Anatoli Blagonravov.

Altogether, up until the end of 1952, the Soviet Union launched nine dogs aboard R-1 rockets, with three flying twice. Improvements continued to be made. For two years, beginning in July 1954, a second series of nine flights would carry other dogs – often more than once – aboard the modified R-1D rocket and, later, the R-1Y variant. Now, instead of landing within the nose cone, each of the space-suited animals was ejected separately as the spent rocket descended. The capsule carrying the dog on the right side would be ejected at around 80 kilometres (50 mi.) altitude. Three seconds later, the parachute would be deployed. The second, left-hand capsule would then eject at around the 45 kilometres (28 mi.) altitude mark, but as a test of the system, the parachute would not be deployed until the capsule was just 3 kilometres from the ground.

In a report filed in December 1956 by Major General Alexei Pokrovsky, director of IAM, the procedures, technology and findings of both series of canine flights from 1951 to 1956 were summarized. The report established that the spacesuits worn by the animals offered sufficient protection throughout the flights; the capsule ejection method functioned well, as did the parachute system, and there appeared to be no behavioural change in either dog post-flight, or any evident physiological harm. Pokrovsky concluded, 'There is no doubt that thanks to the collective efforts of the various branches of science, thanks to the efforts of scientists of all countries, it will be possible to realize manned rocket flight in view of the studies of the upper layers of the atmosphere.'[5]

Five further flights would be carried out during 1957, with the dogs now carried aloft on the more powerful R-2A rocket, which meant they experienced several minutes of weightlessness before the carrier rocket dipped back towards the ground. Unlike in the second phase of flights, the dogs once again landed by parachute in their protective capsule, still contained within the rocket's nose cone. On the second such flight, on 24 May 1957, the R-2A rocket carrying dogs Ryzhaya (Ginger) and Dzhoyna (Johnny) reached an altitude of 212 kilometres (132 mi.), but sadly the animals died after the capsule depressurized.

ON 4 OCTOBER 1957 the world awoke to a new era in global history with the news that the Soviet Union had successfully launched a satellite named Sputnik (Fellow Traveller) into orbit, carried aloft by a far more powerful R-7 rocket. The satellite itself was the size of a beach ball, 58 centimetres (23 in.) in diameter, and it weighed just under 84 kilograms (184 lb). Sputnik circled the globe every 98 minutes on an elliptical orbit, transmitting a constant *beep-beep* sound to tuned-in listeners below and ushering in the beginning of the space age.

The Soviet premier Nikita Khrushchev was understandably overjoyed with the worldwide propaganda value realized by the satellite and sought out Chief Designer Sergei Korolev. He explained that he wanted a second Sputnik satellite launched in time for the anniversary of the Russian Revolution on 7 November – just a month away. Korolev knew his OKB-1 design and manufacturing team was up to the task, and he even promised Khrushchev an added bonus – Sputnik 2 would be launched with a small dog on board.

Korolev gathered the OKB design team together and laid out his plans. He emphasized that this was a monumental undertaking and

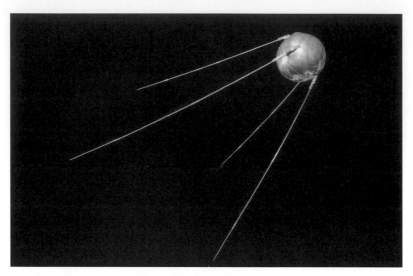

The first Earth-orbiting artificial satellite, Sputnik.

many shortcuts would have to be taken in the name of expediting the launch. As work got underway, design engineers drew up rough plans and passed these directly to workers on the factory floor, who translated them into hardware. From the outset, it was known that there was no way of safely returning this new satellite. The existing parachute systems simply could not cope with the sudden massive drag of hurtling back from orbit at a high speed, so all preparations were completed knowing that it would be a one-way journey for the selected canine passenger.

The animal would travel in a relatively simple pressurized cabin similar to those used on earlier ballistic biomedical flights, launched aboard a stripped-down R-7. This meant the elimination of the satellite separation mechanism, so the satellite would go into orbit while still attached to the upper stage of the rocket. Previous life-support systems for the dogs had included an automatic feeder, but this also came under discussion. As OKB-1 engineer Arkady Ostashov later recalled, the flight was only planned to last seven days. 'We suggested that he [Korolev] save a few kilograms by designing the feeder for one meal only, since we were mainly interested in knowing whether the dog would be able to eat at all.'[6] Korolev agreed, and a simple tin box containing all of the dog's food would be fitted into the pressurized capsule. The lid could be electrically opened by a ground-control operator, giving the dog access to what would be her one and only meal.

Meanwhile, at the IAM, Oleg Gazenko and his biomedical team had been handed the task of selecting the best dog for this historic but eventually fatal mission. Following numerous tests, they finally settled on a 6-kilogram (13 lb) two-year-old mixed-breed female with a suitably calm disposition, casually known to her trainers as Kudryavka (Little Curly). She had been one of the strays plucked earlier from the streets of Moscow. Another dog, named Albina (Whitey), would serve as the back-up if a change became necessary.

On 3 November 1957, just thirty days after the launch of the first satellite, Sputnik 2 lifted off from the newly founded Baikonur launch complex, east of the Aral Sea in the desert steppes of Kazakhstan. Strapped into her capsule within the nose cone was the dog formerly known as Kudryavka, now bearing the name Laika (Barker) after her breed type – and the fact that she had quite a loud bark.

Early monitoring indicated that Laika was understandably stressed during the early launch phase, and her respiration rate increased along with the g-forces of acceleration. Once the strap-on booster engines had been expended and jettisoned, the sustainer-stage engine at the base of the rocket kicked in, maintaining the velocity of the ascending vehicle. The readings soon returned to normal, although the transition to weightlessness seemed to puzzle Laika and take her by surprise. Once orbit was achieved, a strong spring ejected the conical protective fairing surrounding the nose cone, which separated into two sections and fell away like rose petals. Following this, two oxygen gas jets backed

Laika, the first dog to be launched into orbit.

Sputnik 2 and the sustainer stage of the rocket away from the fairing sections to avoid a possible collision. Despite all that was going on, the indications were that Laika remained calm.

As the flight progressed, however, the temperature in Laika's cabin grew steadily warmer, both from Laika's body and from the fierce heat of the Sun beating on the surface of the metal nose cone. The decision to keep the satellite attached to the sustainer stage – separating them was a pointless technical exercise – may also have exacerbated the problem. The insulation and cooling fans were proving useless and the little dog was becoming increasingly stressed and panicky. Eventually the temperature in her cabin climbed beyond 40°C (over 100°F). Laika's biometric signals began to falter, and sometime after the fourth orbit (five to seven hours into the flight), they failed completely. It is believed that Laika succumbed to heat prostration and died.

Meanwhile, official news bulletins ignored this sad development for propaganda purposes and continued to maintain the lie that Laika was doing well and her orbital flight was progressing as planned. The Soviets had admitted soon after Sputnik-2 achieved orbit that the spacecraft would not return to Earth, as the technology needed for a safe return had not yet been developed. A worldwide outcry erupted over the fact that Laika – this brave little 'Muttnik' as she was known in the West – was doomed from the outset and would eventually die in space. Sputnik 2 would remain in orbit for a total of 162 days, completing 2,570 Earth orbits before re-entering and burning up in Earth's atmosphere in 1958.

The truth behind exactly when Laika died was finally revealed in 2002 when Soviet scientist Dmitri Malashenkov addressed the World Space Congress in Houston, Texas. For more than four decades, the official version had been that Laika died painlessly about a week into her mission.[7]

Following the orbital flight of Laika aboard Sputnik 2, Korolev and his design team continued to assist in sending further dogs on suborbital flights, reaping further vital information and biomedical data. There would be five such rocket flights in 1958, and two each in 1959 and 1960. One such orbital flight resulted in the death of two dogs, named Zhulka (Cheater) and Knopka (Button), on 31 October 1958, following a parachute failure.

WITH SEVERAL YEARS of data gathered from numerous biomedical animal flights, together with rocket and spacecraft systems checks, ejection seat and parachute tests, the Soviet space chiefs decided it was time to raise the stakes and proceed with plans to send the first human into space.

Before the first piloted Soviet mission could take place, however, several precursory test flights of the Vostok (East) spacecraft were carried out, with mixed results. The first flight (designated Sputnik 4) was launched on 15 May 1960. It was never intended that this minimal Vostok craft would be recovered, so many later features such as the thermal (heat shield) coating, parachute system and ejection seat were not included. The first phase of the launch went well, but a problem in the guidance system meant that the spacecraft was positioned in the wrong orientation for retrofire, thrusting it into a higher-than-expected orbit. The descent module, its orbit slowly decaying under the influence of microgravity, eventually re-entered and burnt up in the atmosphere on 6 September 1962.

Since the first test flight of a Vostok-type spacecraft had achieved a reasonable amount of success, the Soviet Union pressed on with the programme. On 28 July 1960 a Vostok capsule was launched under the designation Korabl-Sputnik 1 (translating to 'vessel satellite' and incorrectly classified in the West as Sputnik 5) with two dogs named Chaika (Seagull) and Lisichka (Foxie) on board. The flight ended abruptly when the Vostok launch vehicle, beset by severe vibrations, disintegrated thirty seconds into the flight. The Vostok nose cone carrying the dogs was explosively detached by ground control and should have landed safely by parachute, but it was jettisoned at too low an altitude, and the

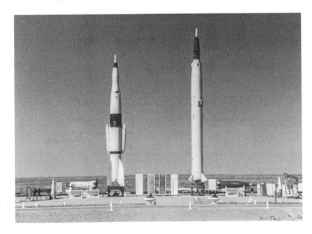

R-2 (left) and R-5 rockets were used to launch the first living creatures into space.

parachute only had time to partially deploy. Chaika and Lisichka were killed on impact. This catastrophic situation would not be a factor on human-piloted flights, as provision had already been made to fit an ejector seat into the later Vostok capsules, and the cosmonauts were in training to eject from their spacecraft before it reached the ground if there was a problem with the spacecraft's parachute system.

The next Vostok (Korabl-Sputnik 2) was launched on 19 August 1960 carrying a menagerie of animals: 28 mice, two rats, a swarm of fruit flies and a number of plants. Also on board were two mixed-breed dogs, Belka (Squirrel) and Strelka (Little Arrow). Owing to the stress associated with launch, the dogs' pulses and breathing rates soared, but these returned to normal once the Vostok craft went into orbit. Their conditions were monitored throughout the flight, and a television camera transmitted pictures of the dogs back to the ground, which soon appeared on the front pages of many newspapers. The animals seemed to cope well with weightlessness, although Belka began barking and became quite agitated during the fourth orbit and was seen to vomit a little.

After seventeen Earth orbits, the Vostok descent module success-fully re-entered and landed by parachute. Belka and Strelka (and their companions) had become the first living creatures to survive an orbital flight. They had spent more than 25 hours in space and travelled around 700,000 kilometres (435,000 mi.) as they circled the globe. The animals were put through a series of medical tests immediately after their flight, which revealed no abnormalities. In fact, six months after the flight, Strelka gave birth to six healthy puppies, one of which Khrushchev

Lisichka (left) and Chaika.

Oleg Gazenko holds Belka and Strelka aloft after their historic orbital flight.

proudly presented to a delighted Caroline Kennedy, daughter of u.s. president John F. Kennedy.[8]

Six months before Belka and Strelka created spaceflight history, a group of twenty elite Soviet Air Force pilots had arrived at a specially built training camp outside of Moscow. They had recently been selected as the Soviet Union's first cadre of cosmonaut trainees, although for most of them, their names would remain a state secret until they flew into space. On 11 October the best-performing half-dozen were formally selected and endorsed as what became known as the Vanguard Six, in order to better concentrate their training on the first Vostok missions. On 18 January 1961 this number was narrowed, with the top three performers chosen and ranked in preferred flight order. These candidates were Yuri Gagarin, Gherman Titov and Grigori Nelyubov (who was later dismissed from the cosmonaut corps for drunken and insubordinate behaviour). Each of them was now hoping to become the world's first person to fly into the cosmos.

Korabl-Sputnik 3 lifted off from the Baikonur launch site atop a Vostok-L carrier rocket on 1 December 1960, this time carrying two dogs named Pchelka (Little Bee) and Mushka (Little Fly). The spacecraft achieved a low Earth orbit, but when it came time to de-orbit after a day in space, the braking system, which was still under development, malfunctioned. The retrorockets failed to shut down when planned, causing the spacecraft to re-enter the atmosphere on completely the wrong trajectory. Rather than allow the spacecraft to fall into foreign hands, the decision was made to detonate an explosive charge within the craft, which disintegrated along with its occupants. Pchelka and Mushka became the last dogs known to have perished during a Soviet space flight.

Three weeks later, on 22 December 1960, yet another Vostok spacecraft was launched from the Baikonur pad, located some 200 kilometres (120 mi.) east of the Aral Sea in Kazakhstan, with another two dogs, Damka (Queen) and Krasavka (Little Beauty), aboard. The five strap-on booster rockets and the core stage of the rocket performed well, but the spacecraft separated from the third-stage rocket while it was still firing and the mission had to be aborted. The Vostok spacecraft subjected the two dogs to a rough and crushing ballistic re-entry. An ejection seat was supposed to jettison the animals to a parachute landing, but it failed to operate, and the dogs landed, still inside the spacecraft, in a deep snowdrift. In freezing conditions, recovery crews finally reached the capsule, disarmed the self-destruct mechanism and retrieved the two dogs, who had managed to survive their ordeal, although the mice accompanying them had died from the cold. If successful, this flight would have been designated Korabl-Sputnik 4, but in light of its being abandoned, it was decided at upper levels to simply refer to it as an anomaly and no name or number was ever applied to that particular mission.

The Soviet Union was moving ever closer to launching its first cosmonaut, and on 9 March 1961 the next test mission, Vostok-3KA, took place, carrying the Korabl-Sputnik 4 designation that had been temporarily assigned to the previous failed flight. This time there was only one canine passenger: a jet-black dog named Chernushka (Blackie). One of the principal goals of this flight was a test of the spacecraft's ejection and parachute systems following re-entry.

Opposite: Dummy cosmonaut 'Ivan Ivanovich'.

IVAN IVANOVICH:
TEST FLIGHT MANNEQUIN

This lifeless space traveler orbited the Earth on March 23, 1961, two weeks before Yuri Gagarin's flight. His mission tested the Vostok spacecraft and its pressure suit, as well as the tracking and recovery operations. The mannequin is named "Ivan Ivanovich," the Russian equivalent of "John Doe." Technicians were so concerned that Ivan's features were so lifelike that they wrote "maket" on the forehead, so anyone finding the mannequin upon landing would not be confused.

Like subsequent Vostok cosmonauts, Ivan was ejected from the spacecraft in an ejection seat after reentering the atmosphere. He parachuted out of the seat and landed near the Ural Mountains city of Izhevsk during a heavy snow storm. Ivan has resided in this spacesuit for over 35 years.

To facilitate this, a space-suited anthropometric mannequin was strapped into the ejection seat, representing a cosmonaut pilot. The mannequin, given the name 'Ivan Ivanovich', was fitted with a series of cages holding forty black mice, forty white mice, guinea pigs, reptiles, plant seeds, human blood samples, human cancer cells, micro-organisms, bacteria and fermentation samples. These were attached to Ivan's chest, stomach and legs. In order to test the communication system, an automatic recording of a choir singing was also fastened inside the mannequin. Altogether, this latest space flight lasted 115 minutes, following which Ivan was successfully ejected above the ground and landed safely by parachute, while Chernushka remained inside the spacecraft, which also came to a parachute landing nearby and was recovered unharmed.

The small dog selected to ride into space on the Korabl-Sputnik 5 mission – the final precursory Vostok flight – had disarmingly soft eyes, but was endowed with a name difficult for Westerners to pronounce: Zvezdochka. The English translation is much easier: Little Star. Rounded up on the dingy side streets of Moscow like so many of her canine contemporaries, Zvezdochka was originally named Udacha, which means 'Luck'. Sometime prior to the launch, the Vanguard Six cosmonauts had visited Baikonur to witness the beginning of the flight. When Gagarin first heard the name of the little street dog, he suggested that Luck was not a good omen or a suitable name for the heroic dog and proposed the change, which was accepted.

Once again, Ivan Ivanovich flew as the substitute pilot of the Vostok spacecraft. This time, however, he had a new head, as the original had been damaged during parachute exercises. Additionally, to prevent any confusion should a civilian come across the inert body of the 'cosmonaut' after landing, the word MAKET (Russian for 'dummy') was painted on the back of his Sokol spacesuit and also written on a card wedged on the inside of his helmet. As before, he was surrounded in the cockpit by a number of small animals, plants and biomedical samples. A further trial of the communication link would take place, in this instance through another recording of the Piatnitsky Choir, followed by a spoken recipe for cabbage soup. It would certainly baffle any Western listeners monitoring the flight.

The mission began on 25 March 1961 and once again was successfully completed after a single Earth orbit, which would be the planned objective of the first piloted mission. The Ivan mannequin was

Four dogs who made spaceflight history. From left: Strelka, Chernushka, Zvezdochka and Belka.

automatically ejected from the spacecraft right on time, landing in 2 metres (7 ft) of snow. Conditions were so difficult that the recovery crews had to wait a full day before they were able to reach the dummy cosmonaut with the help of villagers using horse-drawn sleds. Zvezdochka and the other animals, meanwhile, safely landed inside their spacecraft in the region of Udmurtia, some 265 kilometres (165 mi.) away. They were all recovered in good condition.[9]

Everything had worked well during the flight and a final decision would soon be made on whether the next Vostok mission should carry the world's first space traveller. That decision was made by the Presidium of the Central Committee of the Communist Party on 3 April. Two days later, the Vanguard Six cosmonauts were flown to the Baikonur Cosmodrome on two aircraft as a precaution against the slim possibility that one plane might not make it. The head of the cosmonaut training detachment, Nikolai Kamanin, was on one aircraft, as was physician Yevgeny Karpov, an assortment of medical personnel and some cameramen who would record the launch event. During the flight, Kamanin was mulling over whom he would recommend as the first to fly. As revealed in his diaries, the top three candidates were Gagarin, Titov

Yuri Gagarin in his Vostok spacecraft.

and Nelyubov. While all three had trained equally well, he felt that at times Nelyubov displayed a little too much arrogance to be assigned to such an historic flight, while Titov had a strong personality, which made him ideally suited to a longer and far more ambitious mission. Gagarin, on the other hand, was extremely competent and personable and would present himself well to the world after the flight as the first Soviet person in space.

WHEN IT BECAME TIME for Yuri Gagarin to leave his family in Moscow, he did not want his wife, Valentina, to become overly stressed on the day he would launch into space, so he told her the lift-off was actually set for 14 April – two days after the planned date. The other five men had been briefed to say the same thing.

Although Gagarin remained as the favoured candidate, the selection of the pilot was still to be finalized, and the six candidates continued trying on spacesuits and conducting tests inside the spacecraft. Then, on 8 April, it became official. Having conferred with Korolev, who agreed entirely with his decision, Kamanin called Gagarin and Titov into his office. Tellingly, Nelyubov was not invited. It was then that Gagarin was informed that he would pilot the first Vostok spacecraft, with a devastated Titov learning that he had been relegated to the back-up position in the event of illness or any unexpected difficulty. The State Commission was informed of this decision, which it officially endorsed.

Three days later, the carrier Vostok-к rocket was trundled out horizontally on railway tracks to the launch pad, where it was carefully hoisted to the vertical. Gagarin's Vostok 3ка No. 3 spacecraft was perched at the top beneath a protective launch shield.

The next day, history would be made.

Yuri Gagarin meeting with Sergei Korolev.

2

'POYEKHALI!'

It was the morning of Wednesday 12 April 1961. Strapped tightly into his ejection couch within the sealed, spherical Vostok 3KA spacecraft perched atop a 30-metre-long (98 ft) rocket, also named Vostok, sat a 27-year-old Soviet Air Force pilot. As the countdown to launch proceeded, he listened to some patriotic music being piped into the spacecraft and would doubtless have been visually checking his minimal control and instrument panels. He may have also been pondering the recent events and experiences that had brought him to this remote launch site on the edge of the Central Asian desert and a powerful rocket that would soon propel him into the many wonders of the cosmos.

The person destined to become the world's first space traveller, Yuri Alekseyevich Gagarin, was born into a family of mostly collective farmers on 9 March 1934. They lived in the village of Klushino, located within the Smolensk Region in the west of the Russian Federation. His father, Alexei Ivanovich, was a carpenter and his mother, Anna Timofeyevna, a milkmaid. Yuri was one of four children growing up in a small log cottage that always smelt of freshly planed wood and apples.

On 22 June 1941 German forces invaded the Soviet Union, bringing with them a dark, tumultuous period of privation, misery, hunger and fear. On 1 October that year, young Yura (as he was known) entered school for the first time, but his classes would only last a month before the school had to be shut down. On 12 October Klushino was overrun by Nazi troops, and the Gagarins, like other families in the village, had their home seized by the occupying forces. They were forced to live in a small, squalid dugout Alexei had built in their back garden.

As a child, Gagarin witnessed many vile atrocities carried out by the Nazis. In small acts of rebellion, along with other local youths, he and his younger brother by two years, Boris, would perform small acts of sabotage. One of these was to temporarily immobilize German tanks by pouring chemicals into their batteries and blocking the vehicles' exhaust pipes with raw potatoes. Caught in the act one time, the Germans callously strung up Boris by the neck from an apple tree before leaving him to die. Fortunately, his mother saw what was happening and managed to cut him down in time. His life had been spared, but he was in a bad way, physically unable to walk for some weeks, and in later years, Boris suffered badly from terrifying nightmares. In the end, a deep and endless despair caused Yuri's brother to take his own life by hanging. The family had also suffered another crushing blow when the two eldest children – Valentin, aged seventeen, and Zoya, fifteen – were transported to Germany for use as forced labour (they survived, but did not return until the war's end).

Eventually, in 1943, Soviet Red Army forces overran and liberated what remained of the battered town of Klushino on Yura's ninth birthday. He was able to resume his early education, albeit in his schoolteacher's house, as the school had been burnt down by the Nazis.

After the war, the Gagarin family moved to the town of Gzhatsk, where Yura continued his secondary studies. In 1951 he completed trade school with honours in the town of Lyubertsy, near Moscow, qualifying as a foundryman. In 1954, while undertaking further studies at an industrial college in Saratov, a little north of Volgograd, an eager Gagarin and some fellow students decided to join a flying club. There was initial disappointment in the amount of manual reading and classroom instruction they had to absorb, and then, remarkably, the students had to complete a parachute jump before being permitted to begin actual hands-on flight instruction. Gagarin's flying lessons continued while he was working as a moulder-smelter in a Saratov foundry. In 1955 he not only graduated from the flying school with high honours but completed his studies at the industrial school, also with honours. However, he was now focusing on a different career, with a burning desire to join the Soviet Air Force.

'I did not become an Air Force pilot by chance,' Gagarin would later declare.

During the war we boys felt powerless. Certainly, we did what we could to hurt the Nazis; we would sprinkle nails and broken glass on the road to puncture the tyres of their cars . . . but when we were older we realised how important our country's security is. And that was what led me to make the choice I did – I dreamed of becoming an Air Force pilot.[1]

Gagarin next attended the military aviation school in Orenburg as an Air Force cadet, where he gained his wings in 1957 and also met his future wife, Valentina Goryacheva, at a dance. Two momentous events occurred later that year. On 4 October, after completing a morning of flight training, Gagarin heard that the Soviet Union had launched an artificial satellite called Sputnik into orbit around the planet. The next day, all of the cadets were devouring information about Sputnik in the newspapers. 'I drew this spaceship in my notebook,' Gagarin would later write, 'and again felt that familiar, somewhat obsessive and not yet recognized urge; that same attraction to space, which I was afraid to acknowledge, even to myself.'[2] Three weeks later, he and Valentina were married.

His training now at an end, Gagarin was assigned to a two-year post with the Northern Fleet of the Soviet Air Force at Luostari Air Base in the Murmansk region of the Arctic Circle, close to the Norwegian border. He was subsequently appointed to Senior Lieutenant Leonid Vasilyev's air defence force. During this time, Yuri and Valentina welcomed the first of their two children in April 1959, a daughter they named Yelena. In July that year, he received a promotion to military pilot 3rd class, and six months later decided to submit a report to his commanding officer indicating his interest in volunteering his services for his nation's space programme.

'In connection with the expansion of space exploration going on in the USSR people may be required for manned spaceflights,' he wrote, 'I request you to take note of my own ardent desire, and should the possibility present itself, to send me for special training.' The letter was passed on in turn to the appropriate Air Force authority by his commanding officer, Lieutenant Colonel Babushkin, who had marked Gagarin's report with own endorsement.[3]

Meanwhile, unbeknown to Gagarin, a top-secret quest had already begun to locate, test and select a number of jet pilots of superior flying ability who were willing to join the nation's first cosmonaut group.

On 6 November 1959, three weeks after being interviewed at length by the medical commission charged with this covert duty, Gagarin was promoted to the rank of senior lieutenant. His next and rapid promotion, to major, would be awarded during his historic space flight.

LAUNCHING THE FIRST PERSON into orbit and returning him safely to Russian soil was not without considerable risk, but the Supreme Soviet leadership was impatient to demonstrate what was proclaimed to be the vast superiority of Soviet space technology. Alarmingly, it dismissed a litany of equipment failures that had created anxiety in many of those close to the Soviet space programme, already concerned over the pressure being exerted to launch the first cosmonaut into the still largely unknown realm of space. Doubts definitely existed as to the reliability of the Vostok carrier rocket. Although a Vostok-class rocket was not involved, a massive launch-pad explosion of a prototype R-16 ICBM at the Baikonur Cosmodrome on 24 October 1960 had resulted in the deaths of a reported 78 bystanders (although other sources place this number as high as 126) and had caused 120 non-fatal injuries. It increased wisely unspoken concerns over what was believed to be the undue haste to launch a cosmonaut, even though this rush was fuelled by competition with the Americans, who were preparing for a piloted suborbital mission.[4]

Other tragic incidents would occur: on 23 March 1961, less than three weeks before Gagarin's launch, the youngest of his nineteen cosmonaut colleagues, Valentin Bondarenko, was severely burnt and died following a catastrophic pressure-chamber fire. Additionally, there were still lingering concerns over the well-being and even survivability of a human being in space. Some scientists feared that the lack of gravity, combined with disorientation and a strong feeling of isolation, might cause a cosmonaut to become panicky and even deranged. The flight would be fully automatic, controlled from the ground, and there was a reluctance to give Gagarin access to – and possibly allow him to override – the controls in case he became badly disoriented. By way of a compromise, they locked a three-digit security code into the vehicle's systems and placed a sealed envelope containing the code in the spacecraft cabin, which Gagarin would only open if a situation arose that required him to take minimal manual control of the vehicle in order to set it on course for re-entry.

Yuri Gagarin wearing his helmet before it was painted with the letters CCCP (USSR).

On the evening of 11 April, Gagarin and his back-up pilot, Gherman Titov, retired early. They were woken the next day at 5:30 a.m., although neither man had actually slept. Both were given thorough medical examinations and found to be in good health. Sensors were then placed on selected areas of their bodies. As they were being suited up, someone in the dressing room casually remarked that after Gagarin had landed in his bright orange spacesuit, he might be mistaken for the pilot of an American spy plane, like Francis Gary Powers, who had been shot down and captured the previous year. A nearby official took the remark seriously and a life-support technician who studied calligraphy as a hobby was recruited to paint 'CCCP' (USSR) in bold, red letters on the front of Gagarin's helmet. This would be revealed some years later, solving the niggling mystery of why the cosmonaut was shown in both a painted and a plain white helmet on launch day. The letters were subsequently applied to the helmets of other cosmonauts.

The two men were then assisted onto the transfer bus that would take them out to the launch pad. The space-suited Titov sat behind Gagarin, and photographs show the other four members of the Vanguard Six cosmonauts on board standing around them in their Air Force uniforms: Grigori Nelyubov, Andrian Nikolayev, Valery Bykovsky and Pavel Popovich. According to an oft-told story, Gagarin is said to have asked the bus driver to pull up halfway to the pad so he could urinate against a front tyre. Whether this story is factual or not, it is now a steadfast tradition that all male cosmonauts observe prior to their own launch. Not to be left out, any females carry a small cup of their own urine that they tip onto a tyre of the bus.

On arrival at the launch pad, Gagarin gave a rehearsed report of readiness to the chairman of the State Commission. Before making his way to the narrow staircase at the base of the launch gantry, Gagarin stopped to embrace several friends and officials and reassured an anxious Sergei Korolev that all would go well.

One of Korolev's leading rocket scientists, Oleg Ivanovsky from OKB-I, was escorting Gagarin that day. He had been one of the team responsible for the construction of the Vostok spacecraft, and as he walked alongside the cosmonaut, he deduced that Gagarin was quite calm and decided to reveal to him the secret three-letter code. As they stopped by the staircase, Ivanovsky leaned close to Gagarin's helmet and whispered the code to him: 1-2-5. Gagarin smiled and said he knew it already – an equally protective instructor had already given him the numbers.

Gagarin and Ivanovsky then climbed up the stairs and after he had waved at the crowd of well-wishers assembled below, Gagarin boarded a compact elevator that would slowly transport him up to the spacecraft level. He was then inserted through the hatch of the Vostok spacecraft and slid feet-first into the ejection seat. After he had been strapped in, Gagarin switched on and checked the radio system using his call sign Kyedr (Cedar); the ground-control response was Zarya (Dawn).

The Vostok rocket and spacecraft being raised on the Baikonur launch pad.

Launch day: Oleg Ivanovsky escorts Yuri Gagarin to his waiting spacecraft.

Ivanovsky then wished Gagarin luck and reached into the cabin. 'I hugged him, shook his hand and gave him a slap on the helmet before getting out,' he reflected in 2007. 'A moment later and the hatch swung closed onto its locks.'[5] Ivanovsky then supervised the sealing of the hatch by his two technical assistants, watching carefully as they tightened each of the 32 securing bolts. Then he heard Korolev in the central control bunker room reporting that a panel light had failed to illuminate, indicating that the hatch was not hermetically sealed. All 32 bolts then had to be loosened and removed so checks could be made and a faulty sensor replaced. Meanwhile, Gagarin was seated quite calmly, humming to the music playing in his spacecraft. Once again, the hatch was closed, the bolts inserted and secured, and this time the all-clear was given.

'I was left alone with the controls,' Gagarin later recalled. 'Now no longer lit by the sunlight outside but by artificial light. Of course I was nervous – only a robot would not have been nervous at such a time and in such a situation.'[6] He talked for a while with Korolev, and then, ten minutes before lift-off, he closed the visor on his helmet and listened to all the interesting pre-launch noises made by the fully fuelled rocket. Three minutes before lift-off, he reported to Korolev that he was ready to go.

At 9:07 a.m. Moscow time, thirty minutes after the hatch had been resecured, the four strap-on engines at the base of Gagarin's Vostok-K rocket ignited and roared into lusty life. Clamps holding the rocket were simultaneously released and the supporting arms swung back and away. The rocket then lifted off the launch pad, rapidly picking up speed as it hurtled into the morning sky. The normally reserved Gagarin, elated and swept up in both the glory and tension of the moment, spontaneously cried out 'Poyekhali!' (Here we go!).

Shortly after lift-off, the ground crew monitoring the flight received a shock when a signal suggested there was a problem with the booster. Fortunately, this turned out to be nothing more than a short interruption in data transmission. By the two-minute mark, the four strap-on boosters had consumed the last of their propellant. They shut down and fell away as scheduled. Next, the payload shroud covering the spacecraft was jettisoned, uncovering the optical viewing porthole (*Vzor*) at Gagarin's feet. Five minutes after lift-off, the rocket's core stage had also been depleted of its propellant. It shut down and was set loose, falling away and leaving just the final stage, which ignited to continue the spacecraft's ascent.

'The flight is continuing well,' Gagarin reported. 'I can see the Earth. The visibility is good. I almost see everything. There's a certain amount of space under cumulus cloud cover. I continue the flight, everything is good.'

Liftoff!: 12 April 1961

Ten minutes after launch, the last-stage rocket fell silent. Another ten seconds went by, and then the Vostok spacecraft was explosively unleashed from the empty rocket and slipped into an elliptical orbit. Gagarin reported further, 'The craft is operating normally. I can see Earth in the view port of the Vzor. Everything is proceeding as planned.'[7] Only after the flight did it emerge that an antenna malfunction had placed Gagarin into a much higher orbit than planned: 327 kilometres (203 mi.) above Earth at its highest point, rather than the originally designated 230 kilometres (143 mi.). Retrorocket firing time was consequently adjusted by ground control.

No public announcement of the launch was made until it was confirmed by officials that the spacecraft had successfully achieved orbital status. The startling news quickly swept around the world. Fifteen minutes after launch, Gagarin radioed that he was over South America. As planned, he drank a little water, ate some puréed meat squeezed from a toothpaste-like tube and, for dessert, consumed a tube of chocolate sauce. Both food items had been specially prepared by the Soviet Academy of Sciences. He reported no difficulty in drinking or swallowing. He had begun writing some of his observations in a notebook until he accidentally let go of his pencil; it drifted out of his reach and was lost. He then had to record the rest of his impressions and sightings on tape. The flight continued to go well as the Vostok craft flew over the south Atlantic, with Gagarin listening to some patriotic music being piped through his cabin to soothe him as his journey continued. The Sun began to rise over the horizon, and all too soon he was preparing for the automatic firing of the liquid-fuelled retrorockets.

A little under an hour into the historic mission, as Gagarin passed over the west coast of Africa, the spacecraft's control system placed it into the required orientation, and soon afterwards, at the 79th minute of the flight, the retrorockets began firing for the scheduled 42 seconds, slowing the craft prior to re-entry. Ten seconds later, again according to plan, a command was transmitted to the spacecraft to initiate the separation of the spacecraft's equipment module from the 2.5-metre-diameter (8 ft) re-entry module containing Gagarin. The separation began, but then an unexpected problem surfaced when a stubborn electrical cable kept the two modules firmly attached to each other.

The tethered spacecraft now began to spin and gyrate erratically as the re-entry continued, and Gagarin's flight was in deep trouble as he crossed over Egypt. If the less protected areas of the descent module were exposed to the ferocious heat of re-entry, both he and his Vostok craft could be incinerated. The temperature in the cockpit soon began to rise alarmingly, but all Gagarin could do was sit tight and hope his spacecraft would stop spinning and right itself. Outside, he could see bright crimson flames lashing the outside of the module. 'I was in a cloud of fire rushing towards Earth,' he would later recall.[8]

As time rushed by, the giddying, uncontrollable rotation was causing Gagarin to lose consciousness. Then, just in time, the cable between the two modules burnt through and separated. Gagarin recalled hearing a loud banging noise as this happened. He soon regained full consciousness and an awareness of what was happening as the descent module rapidly settled into the correct orientation for re-entry. It had been a very close call. As his spacecraft plunged ever deeper into the atmosphere, the wild gyrations quickly dampened, and Gagarin was able to concentrate on his upcoming landing procedures. He was able to report he was okay as the g-forces began to build on his body, pressing him ever deeper into his seat.

As the plummeting Vostok descent module reached just over 7,000 metres (23,000 ft) above the Saratov Region of the Soviet Union, an automatic command deployed the spacecraft's landing parachute. It snapped open, and then a following command blew off the Vostok hatch behind Gagarin's head. Two seconds later, still tightly strapped into his ejection seat, Gagarin was jettisoned outwards, head first. Soon after the ejection seat fell away, his primary parachute deployed almost immediately, and to his immense relief blossomed above him. A potentially hazardous situation arose when his reserve parachute also deployed, but fortunately it did not fully open, as it could have fatally tangled both parachutes. Six minutes after ejecting, the world's first spacefarer touched down in a freshly ploughed field close to the village of Smelovka. The round descent module, meanwhile, had descended under its own parachute and thumped down into a field a little over 2 kilometres away. As he cleared himself of his parachute harness, a woman and a young girl leading a calf nervously approached Gagarin and asked if he had come from space. 'As a matter of fact,' he replied, 'I have!'[9]

His descent trajectory had been monitored, so even though he had landed well away from the planned area, recovery units were quickly on the scene, and they reported to ground control that Gagarin was

The world's first person to fly into space: Yuri Alekseyevich Gagarin.

safe and in apparent good health and spirits. Soon afterwards, a dramatic radio announcement confirmed to an overjoyed nation that Major Yuri Alekseyevich Gagarin (he was promoted twice during his flight) had 'landed safely in a prearranged area of the USSR at 10:55 a.m. Moscow time, following an epic orbital journey lasting 108 minutes'. His now-charred Vostok spacecraft had reached a speed exceeding 27,000 kmph (around 17,000 mph), some three times faster than any other person in history.

The exact spot where he had touched down that day was marked with a wooden post (later to be replaced by a permanent stone obelisk). It bore a simple, temporary plaque on which was written, 'Do not remove! 12.04.1961, 10:55 Moscow time.' The landing, incidentally, was actually two minutes earlier, according to official records.

Following his post-flight debriefing and meeting with various dignitaries, Gagarin was eventually flown back to Moscow two days after his history-making flight, amid a vast outpouring of jubilation, relief and pride in his accomplishment. As he and a beaming Premier Khrushchev kissed, embraced and waved to an enormous cheering crowd that had quickly gathered in Moscow's Red Square, the propaganda machine went into full swing, broadcasting news of this historic feat to all corners of a stunned world. Gagarin would later set off on an extensive world tour, with crowds of people eagerly flocking to see and cheer the world's first person to travel into space – even if he was a Russian.

At first, the Soviet Union announced that Gagarin had landed inside his spacecraft, and it would be some time before it was revealed that he had landed separately under his own parachute. In order to achieve any new official flight record under strict rules laid down by the governing body, the FAI (Fédération Aéronautique Internationale), a pilot was required to take off and land in his aircraft (or, by extension, his spacecraft), hence the deception. It would be some thirty years after his historic orbital flight until information on Gagarin's dramatic and near-fatal re-entry was fully revealed.

A totally overwhelmed Gagarin was showered with awards and honours; he received the Tsiolkovsky Gold Medal of the Soviet Academy of Sciences, the Gold Medal of the British Interplanetary Society and two awards from the FAI. He was later appointed commander of the cosmonaut team and in 1964 became deputy director of Zvezdny Gorodok (Star City), the cosmonaut training centre located outside of Moscow. Among other duties at the centre, he oversaw the selection and training of the first group of five women cosmonauts. He also served as a ground communicator for four later space flights in the Vostok and Voskhod programmes.

In 1967 Gagarin served as back-up pilot for the ill-fated Soyuz-1 mission, in which his cosmonaut colleague Vladimir Komarov perished following a trouble-plagued aborted space flight and a fatal parachute failure. Try as he might, Yuri Gagarin would never fly into space again.

3

VOSTOK FLIGHTS CONTINUE

On 5 May 1961, just three weeks after Gagarin's orbital odyssey, the USA's National Aeronautics and Space Administration (NASA) launched America's first astronaut, Alan Shepard, on a fifteen-minute ballistic space flight. Two months later, astronaut Virgil (Gus) Grissom replicated Shepard's mission with the space agency's second suborbital Mercury flight, which lifted off from Cape Canaveral on 21 July. Much was being made of comparisons between the space programmes of the two competing space nations, and the fact that Gagarin had completed an orbit of Earth, while the Americans could only manage two seemingly lesser suborbital missions. Both Mercury flights had reached beyond the recognized boundaries of space, but returned to an ocean splashdown just a few minutes later. The contrast between the achievements of both nations seemed to indicate to Americans that the secretive Soviet Union enjoyed a technical superiority, both in rocket power and spacecraft design.

Dismayed at these constant comparisons, NASA decided to accelerate its space programme, bringing piloted suborbital test flights to an immediate end and announcing that the next space flight, near the end of 1961, would launch an American astronaut on a three-orbit flight around the world. Armed with this useful information and NASA's publicly announced timetable, Korolev and his fellow designers began setting out plans for a second orbital mission that would precede the u.s. space shot. Originally planned as a similar three-orbit mission, an audacious upgrading of objectives would see the flight of Vostok-2 now extended to a full day and seventeen orbits, which would create a

Cosmonaut Gherman Titov undertaking parachute training for his Vostok-2 mission.

substantially more impressive endurance record than anything NASA was planning. The Soviet cosmonaut assigned to the Vostok-2 mission was Yuri Gagarin's one-time back-up, Major Gherman Titov.

Gherman Stepanovich Titov was born on 11 September 1935 at Verkhneye Zhilino in the Altai Krai region of western Siberia. His father was a language teacher who instilled in his son a love of music, poetry and literature – particularly that of Aleksandr Pushkin – which would remain with him for the rest of his life. At a young age, he was filled with admiration for an uncle who had flown combat missions during the First World War. It was not just his war stories that enthralled young Gherman, but his uniform and shiny black boots. He decided he wanted to be a fighter pilot one day, just like his uncle.

In July 1953 the seventeen-year-old Titov was accepted for aviation cadet training and entered the 9th Military Air School in Kustanai, Kazakhstan. Two years later, he transferred to the Stalingrad Higher Air Force School. Following his graduation as a jet fighter pilot in September 1957, Titov served as a combat pilot with two different Air Guard regiments of the 41st Air Division while based at a village called Siverskaya in the Leningrad military district. During this time he met and had a whirlwind romance with Tamara Vasilyevna Cherkas, whom he later married.[1]

It was in 1959 that Titov underwent a mysterious selection process for what he was told was 'a new field of aviation'. The following year, once he had passed scrutiny and all the physical and psychological tests, he became a member of the Soviet Union's first intake of twenty cosmonaut trainees. He worked hard and later emerged as one of six stand-out candidates (the Vanguard Six) in competition for the first piloted space mission aboard a Vostok spacecraft. He was deeply disappointed, even a little angry, not to have been selected to make the first flight, which went to Yuri Gagarin, but he was assessed during training as being a little too aloof and intractable and temperamentally more suited to a longer space mission. Gagarin, on the other hand, possessed a ready smile and an engaging, humble personality, perfect for the Soviet propaganda machine. When the time came, it was decided to award Titov the planned day-long orbital flight, with the transmission code name of Orel (Eagle).

On the morning of 6 August 1961, Major Titov and his back-up, Andrian Nikolayev, were woken early, had a light breakfast and showered before being given a final medical examination. Next, they had sensors fitted to their bodies and were helped to don their heavy spacesuits and helmets. Both men then officially affirmed their intention to carry out the mission before climbing aboard the blue bus that would carry them out to Launch Site No. 1.

There has never been any written mention of him stopping the bus to urinate on a wheel, so the veracity of this space-age superstition

– said to have begun with Gagarin requesting a comfort stop on the way to the pad – will likely never be resolved, but it remains a tradition for all launch-bound crews. On arrival at the pad, Titov waved and made his goodbyes as he walked to the base of the Vostok rocket, where he and Nikolayev performed a 'space helmet kiss'.

THE FLIGHT OF VOSTOK-2 began from the same launch pad as Yuri Gagarin had done four months earlier. The multistage rocket lifted off just three seconds prior to 9 a.m. Moscow time; within minutes, Titov reported that the spacecraft had achieved close to the planned orbit. Once he had tended to various duties, he telegraphed greetings to Premier Nikita Khrushchev.

The spacecraft had completed its first orbit at 10:38 a.m. With all in order and the cosmonaut safe and well, the successful launch was announced over Moscow radio twelve minutes later. The Soviet news agency TASS reported that 'the aims of the flight are to study the effects on the human organism of prolonged orbital flight and the consequent landing on the Earth's surface, and to study man's working capacity during a sustained state of weightlessness.'[2]

Kept under constant television observation throughout his flight, Titov – like Gagarin – made observations through the spacecraft's central porthole, located beneath his feet, but unlike Gagarin, he had time to take photographs and even filmed some short scenes of the planet passing beneath Vostok-2. He also communicated regularly with Soviet ground control. He manoeuvred Vostok-2 twice using manual controls, although for the most part he remained a passive passenger, with the flight monitored and controlled from the ground. Two hours after achieving orbit, he was meant to consume his first meal, in paste form from tubes, but he had no appetite and only drank some blackcurrant juice.

Titov was enjoying the sight of the blue planet below him, and on the sixth orbit was so overcome with excitement that he cried out, 'I am eagle! I am eagle!' He was supposed to have supper around that time before settling back for a designated sleep period, but once again he had little appetite and even though he managed to consume some paste food from a tube, it only served to increase his feeling of nausea. At one stage he even vomited a little.

Just after falling asleep – which, under the circumstances, he found surprisingly easy – he woke up to find his arms floating in front of him. After tucking them under a safety belt he fell asleep again. He woke a

This depiction of a Vostok spacecraft in orbit shows how Gherman Titov would have been positioned inside the spherical descent module.

few times, but drifted back to sleep until 2:37 a.m., having achieved a remarkable eight hours and seven minutes of sleep as the world slipped by beneath him. Despite sleeping comfortably, he was still feeling a little nauseated but managed to consume a small amount of food for his breakfast.

The flight of Vostok-2 came to an end when the descent module re-entered and touched down by parachute at Krasny Kut, 65 kilometres (40 mi.) east of Saratov on the Volga River, with Titov also landing by parachute 5 kilometres (3 mi.) away. Once again, there was a problem when the reserve parachute dropped out and – as with Gagarin – threatened to tangle with the main chute. Then he found himself floating straight towards a railway line, just managing to drift over the top of a Moscow-bound train.[3]

Altogether, Titov's mission had lasted 25 hours and 18 minutes, and he had completed just under eighteen Earth orbits for a total distance of 703,143 kilometres (436,912 mi.) from launch pad to landing site. His flight, which caught Western observers by surprise, came as yet another blow to the morale of the American people, bringing more unrelenting pressure to bear on NASA, still months away from sending Mercury astronaut John Glenn on what now seemed by comparison a less impressive mission.

On 8 August, the day after his return, Titov gave a press conference for the Soviet media in which he reported that 'weightlessness does not interfere with man's capacity for work.' He added that Vostok-2, which he said he had controlled manually for about two hours, 'was a very smart machine . . . very easy to guide; I could turn it any way I wished. I could steer it in any direction needed, and I could land it wherever I wanted . . . I felt like a real pilot.'[4]

At another press conference three days later, Vladimir Yazdovsky from Moscow's IAM and Oleg Gazenko from the Soviet Academy of Sciences disclosed that Titov had actually suffered from 'unpleasant sensations', possibly due, they said, to weightlessness. They reported that he had felt sick for a 'considerable portion of the flight', and this illness, which 'resembled seasickness', caused a 'disturbance of spatial analysis' (a feeling of disorientation) and loss of balance but was minor enough that 'a sufficient level of working capacities' was retained during the entire flight. The two scientists agreed that weightlessness caused 'a definite instability of central nervous systems reactions' in Titov.[5] Even though it might have been an illness isolated to rookie space pilot Titov, it caused enough concern at the time that investigations were carried out by leading physicians, and it was resolved that medical precautions would be in place ahead of future flights.

According to Chuck Oman, a neuroscientist at the Massachusetts Institute of Technology (MIT),

> They didn't know what they were dealing with. At first they wondered if it was triggered by some central nervous system reaction to fluid shift in the body. Later, we found that wasn't true. But when you go to orbit, you do change the rules. Humans are fundamentally flatlanders. Even though you're not standing on the surface, the brain wants you to have one.[6]

The mystery illness that had struck Titov would later be identified as space adaptation syndrome, or SAS, which is unpredictable and still affects around half of the space travellers from all nations during the first few days of their flight. The cause of SAS is still not fully understood, although recent experiments seem to indicate it is related to our inner ear and an inability to distinguish up from down while in orbit, a sensory confusion – more pronounced in larger spacecraft cabins – that results in dizziness and nausea. The biggest concern for astronauts

Titov chats with President John F. Kennedy and astronaut John Glenn in Washington, DC, 1962.

assigned to a spacewalk, or EVA (extra-vehicular activity), is the knowledge that a sudden bout of SAS could cause them to vomit inside their helmet, likely resulting in a terrible choking death.

By August 1961 only four people had travelled into space, with Gagarin and Titov the sole spacefarers at that time involved in orbital missions. Gagarin had not reported any feelings of nausea during his flight, so Soviet scientists suspected that Titov's illness might be due to a physiological problem he had in adapting to weightlessness. As a result, even though he was keen to continue flying into the cosmos, Titov was effectively grounded, and despite receiving a tentative 1964 assignment to an early Soyuz mission, he would never venture into space again.

On 3 May 1962 Gherman Titov and John Glenn (who had flown his three-orbit Mercury mission ten weeks earlier) became the first rival spacemen ever to meet and exchange views during a COSPAR (Committee on Space Research) gathering in Washington, DC. After they and their wives had toured the capital and paid a visit to President Kennedy at the White House, the two space travellers held a news conference at the Great Hall of the National Academy of Sciences. For about 45 minutes, newspaper correspondents relentlessly pressed Titov for technical details of his Vostok-2 spacecraft and carrier rocket, but he was circumspect in responding to those questions. He did confirm, however, that – as widely suspected in the West – he had parachuted to the ground after being ejected from his spacecraft after it had re-entered the atmosphere. This basically confirmed that Yuri Gagarin had likewise parachuted separate to his Vostok craft, instead of landing in it as falsely reported in the Soviet media at the time.[7]

Following his first and only space flight, Titov (like Gagarin before him) was named a 'Hero of the Soviet Union' and received a second Order of Lenin. In 1962 he became a deputy of the Supreme Soviet, a position he held until 1970. In the latter part of the 1960s, he became involved as lead cosmonaut in the Soviet Union's Spiral spaceplane project but lost any chance of a second space flight when, following the death of national hero Gagarin, he was officially grounded from flying into space in March 1968. That same year, he graduated from the Zhukovsky Air Force Engineering Academy. In 1975 Titov achieved the rank of major general. Following the demise of the Soviet Union in 1991, he resigned from the Air Force with the rank of colonel general and entered politics. Four years later, he was elected to the State Duma, the lower house of the Russian parliament, as a member of the Communist Party. He tried, but failed, to achieve a second term in 1999.

Colonel General Gherman Titov died on Wednesday 20 September 2000 while taking a sauna in his Moscow apartment. His death at 65 was officially ascribed to heart disease, despite early but erroneous reports that he had succumbed to carbon monoxide poisoning. Quite remarkably, six decades after his first and only space flight, Gherman Titov still holds the record – at 25 years of age – as the youngest person ever to fly into orbit.

THE FLIGHT SUCCESSION for the Vanguard Six cosmonauts would continue with Vostok-3. Having served back-up duties for Gherman Titov on the Vostok-2 mission, Major Andrian Grigoryevich Nikolayev, a 32-year-old bachelor from the village of Shorsheli in the Chuvash region, was assigned to the Soviet Union's third solo flight in the Vostok series. He was actually named to the mission on 20 February 1962 – the same date as the launch of u.s. astronaut John Glenn on his three-orbit Mercury flight. Unlike Glenn, however, the identity of the third Soviet cosmonaut to fly would remain a state secret until he was in orbit. In making his selection, General Nikolai Kamanin, head of cosmonaut training, also named 31-year-old Lieutenent Colonel Pavel Romanovich Popovich to the following Vostok-4 mission. Their back-ups would be the last remaining members of the Vanguard Six: Valery Bykovsky and Grigori Nelyubov.

In the United States, there had been a growing unease in the corridors of NASA as the months slipped by without an expected Soviet space spectacular. In May 1962 Scott Carpenter completed NASA's second three-orbit Mercury mission, and the space agency was gearing up for a much longer flight with Wally Schirra on board in October.

Following the public euphoria that ensued after the epic space flights of Gagarin and Titov, Soviet premier Nikita Khrushchev was growing increasingly impatient as his nation's space programme seemed to be losing its previous advantage, while American astronauts and their successful missions were capturing all the headlines. One of the major reasons for the delay in sending cosmonauts on further flights was the priority being given to the covert Zenit military photoreconnaissance satellite programme, for which Vostok rockets carrying the satellites would be launched from the sole suitable launch pad at Baikonur.

The first Zenit launch began well enough on 11 December 1961, but a fault in the Vostok rocket's third stage triggered an explosive destruction of the Zenit satellite. The second launch on 26 April 1962 was successful, but the photographic results were poor owing to a problem in the orientation system of the satellite. The third launch, on 1 June 1962, ended in complete disaster when one of the rocket's strap-on boosters shut down and became dislodged during lift-off. It fell back onto the launch pad and exploded, causing extensive damage to the pad, while the now wayward rocket arced and plummeted to the ground, crashing and exploding 300 metres (985 ft) away. This damage would further delay any attempts to launch Vostoks 3 and 4.[8]

Annoyed at these delays, Khrushchev consulted Marshal Dmitry Ustinov, deputy chairman of the Central Committee of the Communist Party, who oversaw the rocket programme. Ustinov agreed with Khrushchev that too much time was passing by in launching further cosmonaut missions. Armed with this validation, Khrushchev met with Sergei Korolev and demanded further human flight spectaculars – and soon. He wanted a programme schedule submitted to him within a few days. Cosmonaut training chief General Nikolai Kamanin was furious with the sudden haste required by this demand. 'Such is the style of our leadership,' he later groused in his diaries. 'They've been doing nothing for almost half a year, and now they ask us to prepare within 10 days an extremely complex mission, the programme for which has not even been agreed upon.'[9] Nevertheless, Korolev and his design team knuckled down to the task once again, coming up with plans for the launch of two piloted flights on consecutive days once the badly damaged Baikonur pad was available for this bold venture. By August, twelve months had passed since Gherman Titov's day-long flight around the world, but Khrushchev's insistence and public speculation would soon come to an end as the Soviet Union bounced back with not one but two crewed Vostok launches in as many days. For Korolev and his team, it was a truly amazing achievement.

On the evening of 7 August, four days before the planned launch of Vostok-3, the State Commission overseeing the cosmonaut flight programme formally announced the crews for both upcoming flights. As expected, Nikolayev was assigned as pilot for Vostok-3, with Valery Bykovsky serving as back-up, and Popovich would pilot Vostok-4. However, there was a shake-up in back-up duties: Grigory Nelyubov had not performed well in centrifuge training and was dropped from the Vanguard Six. Vladimir Komarov was brought into the squad to act as back-up pilot for Popovich, while Boris Volynov would assist in a support role, ready to take on any primary or back-up position if required.

AT 11:30 A.M. MOSCOW TIME on 11 August 1962, Andrian Nikolayev became the third cosmonaut and fifth space traveller to orbit the planet when he was launched from the repaired Baikonur pad. This time an official announcement of the Vostok-3 launch over Moscow radio was delayed until 1 hour and 26 minutes into the mission. It stated that the cosmonaut was in good health.

News of the launch broke like a welcome thunderclap across the Soviet Union; work places erupted in joyous scenes, while radios in parked cars were turned up, with news reports attracting crowds of eager listeners. Despite the jubilation, the TASS news agency announced this latest mission in a very low-key manner, merely stating, 'The purpose of the flight is to obtain additional data on the effects of space flight conditions on the human body.' There was much speculation that this latest space flight would go beyond the seventeen orbits achieved by Gherman Titov a year earlier. In fact, the Vostok-3 mission was intended to last four days, if all went well, with a target of 64 Earth orbits.

The Vostok 'space twins' Pavel Popovich (left) and Andrian Nikolayev.

A second announcement the next day, 12 August, caught by surprise not only the citizens of the Soviet Union, but the entire world. It was devastating news for NASA and a worrisome development for the American population. A little less than 24 hours after the launch of Andrian Nikolayev aboard Vostok-3, another rocket had pierced the skies, carrying Pavel Popovich into orbit aboard Vostok-4. There was immediate speculation of a possible rendezvous and link-up in orbit, which sent NASA into a mild panic, as this, by comparison, would set back their own pre-publicized space efforts in the eyes of the American public. Fortunately for the U.S. space agency, the Vostok spacecraft had no means of manoeuvring, which meant that despite the accuracy and timing of the Vostok-4 launch, the closest the two spacecraft could achieve was one occasion when they were passing within 6 kilometres (3.7 mi.) of each other. Nikolayev reported that he could see the other spacecraft through his porthole.

The dual flight created immediate headlines around the world, with many displaying the arresting banner 'Two Russians in Space'. In reality, this was incorrect, as neither man was actually a Russian citizen. Nikolayev was a Chuvash, a Turkic-speaking ethnic minority group descended from the ancient Bulgars, while Popovich was proudly Ukrainian.[10]

As well as supplied tubes of food paste, the two cosmonauts were able to consume plastic-wrapped sandwiches and pastries, roast veal, chicken fillets and fresh fruit. In one planned experiment, both men unstrapped themselves from their seats several times and floated within their respective spacecraft cabins. Both reported no problems working and eating while floating freely in a state of weightlessness, which came as a relief to physicians following illness concerns raised after Titov's mission. Their tandem flight also notched up another notable first when they succeeded in broadcasting television images back to Earth.

On 15 August, according to schedule, the two spacecraft's braking systems fired and they re-entered the atmosphere. In the final stages of their descent over the desert steppe area of central Kazakhstan, both cosmonauts were ejected from their spacecraft. Nikolayev touched down by parachute north of Lake Balkhash at 9:55 a.m. Moscow time, and Popovich – after battling strong winds, which gave him a rough landing – hit the ground six minutes later near Atas, south of Karaganda and some 305 kilometres (190 mi.) to the west of his 'space twin'. Nikolayev's flight aboard Vostok-3 had lasted 94 hours and 28 minutes, completing 64 orbits, while Popovich ran up 70 hours and 56 minutes over 48 orbits.

PREMIER NIKITA KHRUSHCHEV was becoming increasingly demanding in his efforts to ensure that the Soviet Union not only maintained the lead in space exploration but demonstrated capabilities that appeared to exceed by a wide margin any available u.s. technology. With the successful completion of the dual flight by Andrian Nikolayev and Pavel Popovich, and basking in the afterglow of worldwide acclaim over the feat, Khrushchev began badgering Sergei Korolev for an even more impressive space spectacular. This time he envisaged three Vostok spacecraft in orbit at the same time, but that demand was rapidly quashed when Korolev pointed out that there were only two Vostok capsules remaining, and it would take considerable time to manufacture a third. Khrushchev then proposed a mission that could be accomplished using the existing spacecraft: knowing that NASA had no plans in place to recruit female astronauts, he wanted a woman to occupy one of the Vostok spacecraft on another tandem flight.

As head of cosmonaut training, General Nikolai Kamanin supported Khrushchev's plan, having already begun a recruitment process to select a number of female cosmonauts. 'We cannot allow that the first woman in space will be an American,' he wrote in his diary. 'This would be an insult to the patriotic feelings of Soviet women.'[11] Khrushchev was delighted with Kamanin's agreement.

Earlier, on 31 December 1962, the Central Committee of the Communist Party had agreed to Kamanin's request to select a second group of cosmonauts, while stipulating that five of them be women, to be trained and flight-ready in 1963. Armed with the understanding that very little flying knowledge would be needed throughout the ground-controlled space flight, a team of recruiters secretly began looking for suitable young women involved in military, acrobatic or sport flying, especially those with advanced parachuting qualifications. On 3 April 1962, following initial and then far more intense screening and testing of candidates, the names of five women were submitted for final approval.

Just a week before the women were accepted, the cosmonaut corps had been rocked by the sacking of three of its original members: Grigori Nelyubov, Ivan Anikeyev and Valentin Filatyev. They were arrested for being drunk and disorderly at a railway station close to the cosmonaut training centre, and Nelyubov in particular had been both belligerent and unapologetic for his rude behaviour to the arresting militia. An example had to be set, and despite Anikeyev and Filatyev being considered innocent victims of Nelyubov's arrogance, all three were dismissed. Nelyubov had once been a member of the Vanguard Six astronauts and a candidate for either Vostok-3 or -4.[12]

The women who would begin cosmonaut training were:

- Tatyana Kuznetsova, 20, a qualified parachutist and holder of several parachuting world records.
- Valentina Ponomaryova, 28, a married mother with a flying and parachuting background.
- Irina Solovyova, 24, holder of many world records while completing 2,200 jumps.
- Valentina Tereshkova, 24, a textile-mill worker and amateur parachutist with over one hundred jumps completed.
- Zhanna Yorkina, 22, also an amateur parachutist.[13]

As all five were civilians involved in a military operation, it was decided to bestow upon them the lowly rank of private in the Soviet Air Force,

Grigori Nelyubov (centre) with Andrian Nikolayev and Yuri Gagarin prior to his dismissal as a future cosmonaut.

which would rise to junior lieutenant once they had satisfactorily completed their cosmonaut training.

Valentina Vladimirovna Tereshkova (known to her family as Valya) was born on 6 March 1937 in the village of Maslennikovo, near the ancient city of Yaroslavl on the River Volga. Her father, Vladimir, a tractor driver, was killed two years later during the Soviet–Finnish War, leaving her mother, Yelena, a weaver at a textile factory, to raise their three children, Ludmila, Valentina and Vladimir. The family later moved into Yaroslavl, where Valentina attended school until she left at the age of sixteen to enter the workforce, first at a tyre factory and then in 1955 as a loom operator at the Krasnyi Perekop (Red Canal) textile mill, where her mother was also employed. She then attended evening classes in order to achieve a diploma from a textile technical institute. In 1960 she graduated from the institute as a cotton-spinning technologist.

Two years earlier, in 1958, Tereshkova had made a life-changing decision to take up parachuting as a hobby and joined the local Yaroslavl Air Sport Club, making her first jump on 21 May 1959. Afterwards she felt exhilarated and vowed to continue. 'I felt I wanted to do it every day,' she later remarked, now spending more and more time at the club. Altogether, she made 126 jumps, which brought her to the attention of the Soviet space chiefs when they were looking for suitable candidates to become the first woman to fly in space.

Tereshkova would recall that on 12 April 1961 she and other members of the club listened intently to reports of a cosmonaut named Yuri Gagarin on a space mission around the world. When she mentioned this to her mother, Yelena responded, 'Now a man has flown in space, it is a woman's turn next.' Little did Valentina know that she would be that woman.[14]

AT PRECISELY 2:58:58 P.M. on 14 June 1963, a Vostok-K rocket lifted off from the Baikonur Site 1 launch pad, carrying Lieutenant Colonel Valery Fyodorovich Bykovsky into the bright blue skies over Kazakhstan. He would use the call sign Yastreb (Hawk). The launch had actually been scheduled to take place two days earlier, but there were mounting concerns over some intense solar flare activity, raising radiation levels above the atmosphere to potentially dangerous levels, so it was decided to try again the following day. The same situation arose on 13 June, causing the planned record eight-day mission to be postponed yet again.

The following morning, the solar activity had subsided, but during the renewed countdown, the spacecraft's hatch had to be re-opened to repair a safety cable beneath the ejection seat. After the hatch had been closed and bolted once again, Bykovsky was informed that a gyroscope had malfunctioned in the third stage of the carrier rocket and needed to be replaced. Another unit was located, tested and installed, and the countdown then proceeded through to launch.

Although the lift-off went mostly according to plan, it was soon realized that Vostok-5 had ended up in a lower-than-expected orbit. This meant it would lose altitude at a faster rate than expected, so an eight-day mission was not achievable. Later in his flight, Bykovsky was informed that his flight would end after four days. Despite this setback, there was frenzied activity back on the launch pad, and Bykovsky would not be alone in orbit much longer.

The Vostok-5 mission created a fresh buzz of excitement across the Soviet Union: not only was another cosmonaut in orbit after a ten-month gap, but there was growing speculation that a 'cosmic rendezvous' might even include the first woman to fly into space.

AT 7:00 A.M. ON THE MORNING of her flight, Tereshkova climbed out of bed at the Baikonur launch centre and did some light exercises, then showered and changed into some casual clothes ready for a specially

The 'First Lady of space', Valentina Tereshkova, prior to her Vostok-6 mission.

prepared breakfast. After their meal, Tereshkova and her back-up, Irina Solovyova, moved to the suiting-up room, where they were both given final medical examinations, had sensors attached to their bodies and then donned their bulky orange spacesuits and white helmets. Soon afterwards, wearing protective overshoes, they made their way onto the waiting blue transfer bus, where Yuri Gagarin and other cosmonauts greeted them before they were all driven out to the launch pad.

After disembarking from the bus, Tereshkova reported to the Chairman of the Commission and formally announced that she was ready for the flight. He wished her a safe journey, following which she mounted the few steps to the elevator level, turned and waved at those assembled below, then climbed into the elevator for the ride up to

where her spacecraft was ready to board. Once there, launch technicians removed the overshoes, assisted her into her seat, strapped her in
securely and then sealed the hatch. With two hours until lift-off,
Tereshkova checked all her radio links and switched on the communication link with Sergei Korolev and Yuri Gagarin in the nearby
underground control blockhouse. As Tereshkova recalled:

> Korolev, named *Zarya* (Dawn), was in constant link-up with
> me, named *Chaika* (Seagull), and Yuri Gagarin, named *Kyedr*
> (Cedar). For me there was no time to reflect on what was
> ahead, with so much work to be done at this stage, waiting
> impatiently for the final words 'lift-off'. Korolev's last words to
> me were, 'Seagull, we wish you a happy flight. We are waiting
> for your return to Earth.'[15]

Valentina Tereshkova's first and only launch into space began two
days after that of Valery Bykovsky, at 12:29:52 on the afternoon of
16 June 1963. The Vostok-6 spacecraft successfully slipped into orbit
after a flawless launch and ascent, beginning what was tentatively
scheduled as a single-day mission that could be extended by up to
three days depending on how Tereshkova was handling the flight. She
had no piloting background to speak of, just some minimal flight
training, and would be strapped into her seat throughout the mission,
while the spacecraft and its systems were essentially controlled from
the ground. Apart from a few nominal tasks, she was little more than
a passenger/observer as Vostok-6 whirled around the planet every
88.3 minutes.

During her first orbit, Vostok-6 came within about 5 kilometres
(3 mi.) of Bykovsky's Vostok-5 spacecraft. The two cosmonauts were in
constant radio contact, but Tereshkova later reported that neither craft
reported seeing the other. After that, the two spacecraft drifted further
and further apart as their flights continued.

As expected, the flight of the first female cosmonaut created extensive press coverage around the world. Not only was Valentina Tereshkova
the first woman to venture into space, but she was young, photogenic
and single. For Khrushchev, she was the perfect choice for the mission,
representing all that was supposedly good and wholesome in the Soviet
Union, where even a young woman raised in a peasant family and
working in a textile factory could achieve greatness.

Bykovsky and Tereshkova with Yuri Gagarin.

Tereshkova, despite her later denials about being unwell on the orbital flight, suffered from headaches and vomiting on the trip. Moreover, her sleep patterns were irregular. Fortunately, her work programme was extremely light and one of her few major tasks was practising a manual reorientation of the spacecraft – necessary in the event of an unexpected manual re-entry. Whether through illness or lack of training, Tereshkova did not perform at all well in carrying out this procedure at first. She obviously found some difficulty trying to manage the technical aspects of her mission, and this, combined with her constant lack of enthusiasm, infuriated Nikolai Kamanin, who wrote a scathing assessment of her performance in his diaries, published many years after the tandem flight.[16]

On the fourth day of Bykovsky's flight, it was decided to bring both spacecraft down the next day. On 19 June Vostok-6 re-entered first, but there was confusion for ground control when Tereshkova failed to report retrofire and the separation of her descent module from the service module, a major fault which had plagued earlier flights. Despite orders to communicate all events relating to this critical phase, she remained strangely silent during this early phase of re-entry.

Nevertheless, the landing procedures went well; Tereshkova was jettisoned from the Vostok-6 descent module as scheduled and landed by parachute at 11:20 a.m. Moscow time, having orbited the Earth 48 times in 2 days, 22 hours and 50 minutes. Everything also went as

Soviet premier Nikita Khrushchev at celebrations for Tereshkova and Bykovsky.

planned for Bykovsky, initiating his re-entry a little under three hours later and landing around 800 kilometres (500 mi.) from Tereshkova at 2:06 p.m. He had completed 81 revolutions of the planet in 4 days, 23 hours and 7 minutes.

Tereshkova would always suggest that she was keen to fly into space again, but her illness and a questionable performance meant this would never happen. Instead, on 3 November 1963, she married fellow cosmonaut Andrian Nikolayev in a Moscow registry office. The following year, their daughter, Yelena, was born, but the marriage – which many suspect was ordered by Khrushchev in order to achieve further publicity for the space programme – did not last and the couple eventually separated. They were divorced in 1982 and Tereshkova would later remarry, this time to a prominent surgeon.

Despite the huge worldwide acclaim harvested by the Soviet Union for sending a woman into space, and the ensuing embarrassment it caused NASA, who had to reluctantly admit they had no immediate plans to enlist women into their astronaut corps, it would be another 21 years before a second Soviet woman was launched on a space mission. Once again it was undoubtedly a propaganda exercise, carried out shortly before NASA launched Sally Ride into space on Space Shuttle *Challenger* as the first female U.S. astronaut.

Valentina Tereshkova remained in the space-training programme for a time as an instructor, and retired from the Soviet Air Force in 1997, having achieved the rank of major general. She would later move into the political arena, holding a number of high-ranking offices before and beyond the fall of the Soviet Union in 1991. She remains politically active to this day while also serving as a globe-travelling goodwill ambassador for her nation.

Valery Bykovsky's flight on the 1963 Vostok-5 mission still holds the record for the longest Earth-orbital flight by any sole spacefarer. In 1967 he was on the verge of flying on the three-man Soyuz-2 joint mission after Soyuz-1 was launched a day earlier with Vladimir Komarov on board. The four cosmonauts were going to attempt to achieve the first docking in space, but their launch was cancelled following the tragic death of Komarov on board the fault-riddled Soyuz-1 spacecraft. Bykovsky's next flight was the eight-day Soyuz-22 mission in September 1976 along with fellow cosmonaut Vladimir Aksyonov. In May 1977 he completed his third and final space flight when he was paired with Sigmund Jähn from the German Democratic Republic on the Soyuz-31 Interkosmos mission to the Salyut-6 space station. Valery Bykovsky passed away in Moscow on 27 March 2019, aged 84, leaving Valentina Tereshkova as the sole surviving Vostok cosmonaut.

Crewed Vostok Missions

Mission	Pilot	Launch Date	Landing Date	No. of Orbits	Duration dd/hh/mm
Vostok *	Yuri Gagarin	12.04.1961	12.04.1961	1	00.01.48
Vostok-2	Gherman Titov	06.08.1961	07.08.1961	17.5	01.01.18
Vostok-3	Andrian Nikolayev	11.08.1962	15.08.1962	64	03.22.28
Vostok-4	Pavel Popovich	12.08.1962	15.08.1962	48	02.22.56
Vostok-5	Valery Bykovsky	14.06.1963	19.06.1963	82	04.23.07
Vostok-6	Valentina Tereshkova	16.06.1963	19.06.1963	48	02.22.50

* In keeping with Soviet space policy, the first Vostok mission had no numerical designation.

4

SOVIET SPECTACULARS
AND A SPACEWALK

In February 1964 a group of cosmonauts found themselves summoned to the OKB-1 (Opytnoye Konstruktorskoye Buro or Experimental Design Bureau) in Moscow by Chief Designer Sergei Korolev. OKB-1 was a closed facility tasked with designing and constructing a number of prototype spacecraft, and the cosmonauts were there to view a new generation of crewed Soviet spacecraft called Voskhod (Sunrise), featuring several technical improvements. Voskhod was 950 kilograms (2,100 lb) heavier than the single-pilot Vostok but externally the same size. Internally, however, it was a little more spacious to allow for dual occupancy.

The cosmonauts listened intently as Korolev explained that after six successful Vostok missions, the programme had come to an end. It was time to move on and keep a step ahead of the Americans, who were already deep into planning for their first two-man Gemini missions the following year. The Soviet space programme could not be permitted to seem inferior to that of NASA in any way, and he reported that plans were in place for multi-crewed flights aboard this interim spacecraft, several of which were then under construction, to fill the gap between the Vostok programme and the readiness of the advanced Soyuz spacecraft, still under development and at least a year or two away.

As outlined by Korolev, planning for the first of four Voskhod missions was well advanced, and though he was against the concept for many reasons – particularly crew safety – he had been instructed (basically ordered) by Nikita Khrushchev to place three cosmonauts into orbit in order to go one better, and earlier, than the two astronauts

NASA would fly on each Gemini mission. The Soviet crew, Korolev explained, would consist of a commanding pilot-cosmonaut, a scientist and a physician, with the launch scheduled for later that year. It would be the only three-man Voskhod flight; the following missions would comfortably revert to two cosmonauts.

The cosmonauts, having studied the interior of spacecraft, were a little bewildered as to how three space-suited crew members could be squeezed inside the cramped spherical craft. The explanation graphically outlined the monumental risks involved in the one-day mission. First of all, driven by logistics, only lightweight flight suits would be issued in lieu of bulky protective spacesuits, and there would be no individual ejection seats, which also meant no parachutes. It would prove a death trap if anything went seriously wrong at any stage of the flight, with no means of escape for the doomed crew. It remains, to this day, the most ill-conceived crewed space mission ever planned and carried out by any nation – all for the sake of propaganda and national prestige.[1]

Colonel Vladimir Komarov was appointed mission commander and would be making his first space flight. In the heavily contested role of scientist-cosmonaut Korolev challenged the names of all other competitors thrust at him and insisted on flying his best spacecraft designer from the OKB, Konstantin Feoktistov, who had helped design

The first Voskhod crew. From left: Konstantin Feoktistov, Colonel Vladimir Komarov and Boris Yegorov.

The Voskhod spacecraft being mated to a Voskhod 11A57 launch rocket.

the Voskhod and other spacecraft. The third place on the crew – the physician – fell to medical doctor Boris Yegorov. Their spacecraft's call sign would be Rubin (Ruby).

A final, precursory test of the Voskhod system and its soft-landing system was carried out on 6 October 1964. In order to maintain strict secrecy, it was only formally identified as the Cosmos-47 satellite. The test was completed satisfactorily.[2]

Six days later, at 10:30 a.m. Moscow time, the first Voskhod crewed mission lifted off from launch pad LC1 at the Baikonur Cosmodrome, situated near the Aral Sea in Kazakhstan – the same pad from which Yuri Gagarin had been launched four years earlier. As with first flights in earlier Soviet space programmes, it did not have a numerical designation. The spacecraft would orbit Earth fifteen times as planned, spending just under a full day in space. During that time, the three cosmonauts were kept busy conducting some minor experiments – difficult but for the most part accomplished in the extremely cramped conditions.

While Komarov was principally involved in piloting and orienting Voskhod, as well as communicating with ground control and updating

Premier Nikita Khrushchev in his Crimean *dacha* via radio phone, Feoktistov was carrying out visual observations, working with a sextant and photographing hundreds of land and oceanic features, as well as conducting experiments on the behaviour of liquids in microgravity. All the while he was carefully monitoring the spacecraft systems. Meanwhile, Yegorov carried out his medical studies, drawing blood samples and monitoring his companions' blood pressure and heart rates. They were allocated a sleep period, but with all that was going on both inside and outside of their spacecraft, this would prove quite difficult.

At the end of their fifteenth orbit, Voskhod re-entered the atmosphere and completed a safe return, touching down in a field on a Soviet state farm.[3] The crew was subsequently flown to Moscow for joyous celebrations in Red Square, where they were surprised to find that Nikita Khrushchev was not there to greet them. Instead, they were met by Leonid Brezhnev and Alexei Kosygin in their first public appearance as the newly installed leaders of the Soviet Union. Shortly after his radio call to the orbiting spacecraft, Khrushchev had been ordered to return to Moscow, where he had been deposed from office.

Sensational headlines announcing the multi-crewed mission erupted all around the world, while NASA officials were openly frustrated and furious at this latest Soviet 'first', which seemed to indicate a larger-than-expected and possibly more advanced spacecraft. Meanwhile, Soviet space industry insiders were appalled at Khrushchev's insistence on flying such a dangerous mission that could have gone horribly wrong as the world watched, and dubbed the entire Voskhod flight a fiasco and a 'space circus'. Neither Feoktistov nor Yegorov

Recovery crews at the landing site of the first Voskhod crewed mission.

Vostok and Voskhod cosmonauts. Top row, from left: Gherman Titov, Yuri Gagarin, Valery Bykovsky and Andrian Nikolayev. Bottom row: Pavel Popovich, Boris Yegorov, Valentina Tereshkova, Vladimir Komarov and Konstantin Feoktistov.

would ever venture into space again, and mission commander Komarov would not survive his second flight aboard the first Soyuz spacecraft.

IT WAS 18 MARCH 1965. For the second time in six months, a crewed Voskhod spacecraft was mounted atop a fully fuelled rocket, once again on launch pad LCI at the Baikonur Cosmodrome. The two cosmonauts strapped into Voskhod-2, who would use the call sign Almaz (Diamond), were mission commander Pavel Belyayev and Alexei Leonov. As the countdown proceeded, they were patiently anticipating the beginning of their one-day mission, during which they would be propelled into orbit, and thirty-year-old Leonov into spaceflight history.

Alexei Arkhipovich Leonov had been selected in the first Soviet intake of twenty cosmonauts back in 1960, and this would be his first space flight. In the lead-up to the mission, his training had been especially difficult and complex, as he would be carrying out a risky but impressive manoeuvre in space that would soon stun the world. For this flight, Voskhod-2 had been specially fitted with an inflatable rubberized airlock, approximately 3 metres (10 ft) long, that would allow a crewmember to exit the craft and perform a spacewalk. 'A sailor must be able to swim in the sea,' Korolev had once stated. 'Likewise, a cosmonaut must be able to swim in outer space.'[4]

The Soviet space chiefs had been in a hurry to complete this latest space mission, knowing that an American astronaut was then in training for what was known in the NASA lexicon as an EVA (extravehicular activity), more commonly called a 'spacewalk'. National pride and technological prestige was at stake, and the Soviets were determined to trump the USA by becoming the first nation to complete this significant space first. With the countdown clock ticking, Leonov was on a tight, rushed training regime as the modified Voskhod he would exit in orbit was undergoing final preparations.

As with the first Voskhod mission, an unmanned test flight was launched into orbit on 22 February 1965 under the designation of Cosmos-57, which included a full simulation of the airlock deployment and inflation. The test was successfully completed and the crewed mission could now proceed.[5]

At 10:00 a.m. Moscow time, the four booster rockets beneath the Voskhod 11A57 rocket ignited and a successful lift-off was achieved. Within minutes, Belyayev and Leonov had slipped into orbit 169 × 473 kilometres (105 × 294 mi.) above Earth, travelling around the planet once every 90.9 minutes. So far, the mission was going as planned.

Prime crew of Voskhod-2: Alexei Leonov (left) and Pavel Belyayev.

Manufacture of the Voskhod-2 spacecraft with the inflatable airlock in position.

Once everything was in readiness and the airlock on the exterior had been inflated, Alexei Leonov prepared himself to become the world's first person to walk in space. In order to verify that his space-walk had actually taken place, he would mount a video camera at the far end of the airlock while wearing a second camera on his chest.

Towards the end of Voskhod-2's first orbit, Leonov passed through the opened spacecraft hatch into the inflated airlock at the end of a 5.35-metre (17.6 ft) umbilical cord. So far all had gone well, and as the spacecraft hatch was re-closed, Leonov opened the end of the airlock into the vacuum of space. Once this had been accomplished, he fixed the camera in place and slid out and away from the glistening spacecraft at the end of his umbilical cord. He was instantly fascinated by what he could see below him.

'The Earth is round!' he recalled exclaiming as he caught his first view of the world. 'Stars were to my left, right, above and below me. The light of the sun was very intense and I felt its warmth on the part of my face that was not protected by a filter.'

Opposite: A close-up of Alexei Leonov during his spacewalk.

'You just can't comprehend it,' he reflected.

Only out there can you feel the greatness – the huge size of all that surrounds us . . . My feeling was that I was a grain of sand. What remains etched in my memory was the extraordinary silence. It was so quiet I could even hear my heart beat. I was surrounded by stars and was floating without much control. I will never forget the moment. I also felt an incredible sense of responsibility.[6]

Over the next few minutes, Leonov enjoyed the sensation of floating in the cosmos and the spectacular views, which he would later recreate in several of his superb oil paintings. After about ten minutes had passed, he knew it was time to wind things up. He only had five more minutes before the spacecraft slipped into Earth's shadow and they would be surrounded by a total darkness. With great reluctance he began to make his way back towards the airlock when he heard Belyayev saying in his ear, 'It's time to come back in.' It was then that Leonov began to realize he could be in serious – even deadly – trouble. He might not be able to get back inside Voskhod-2.

Until that time, no one had foreseen what might happen to a spacesuit in the vacuum of space. The difference in air pressure had caused Leonov's suit to slowly inflate and it had begun to balloon out of shape. As the suit expanded and stiffened, he felt his hands slide out of his gloves and his feet slip out of his boots. 'My suit was becoming deformed . . . [it] felt loose around my body,' he later reflected. 'I had to do something. I couldn't pull myself back using the cord. And what's more, with this misshapen suit, it would be impossible to fit through the airlock.'[7]

The airlock, he knew, was only 1.2 metres (47 in.) in diameter. Now drenched in perspiration, which was fogging up the inside of his visor and making visibility increasingly difficult, Leonov opted to go radio silent, just in case any Americans were covertly listening in. Back in Moscow, the crew's transmissions, which had been broadcast over state radio and television, suddenly ceased, replaced by constant, ominous replays of Mozart's *Requiem.* Leonov knew he had to act, and quickly.

Going totally against strict training instructions, Leonov opened a valve in his suit, allowing precious oxygen to flow out into space, and the suit began to deflate. He only had 60 litres of oxygen for

The actual Voskhod-2 spacecraft on display at the RSC Energia Museum, Moscow, with a model of the extended inflatable airlock and a mannequin Leonov at the opening.

ventilation, and losing a good percentage of it carried the risk of decompression sickness – 'the bends'. He began to feel the warning signs of pins and needles in his hands and feet. 'I was entering the danger zone; I knew this could be fatal.'

Normal procedure called for a feet-first re-entry into the airlock. As soon as he felt the suit had deflated sufficiently, he thrust himself head first into the opening, dragging himself in. He then had to somehow turn around in order to pull in the umbilical tether. 'It was the most difficult thing,' he recalled. 'I'm in this suit and I had to turn around in the airlock. But with the perspiration, I couldn't see anything.'[8]

Eventually he succeeded; the opening to the airlock was closed tight, and the spacecraft hatch was opened when the pressure had equalized. Leonov crawled through into the cabin, slumping into his seat beside Belyayev, shaking violently and totally spent as he lifted his dripping visor.

As Leonov recovered, Belyayev initiated the release of the airlock. The moment he did, the crew had another serious situation as Voskhod-2 began to spin violently. G-forces rapidly escalated and blood vessels in both cosmonauts' eyes ruptured. As if that were not enough to contend with, a mechanical problem sent oxygen levels in the cabin climbing rapidly, which created a highly flammable situation. According to Leonov, 'A spark would have caused an explosion and we would have been vaporized.'[9] Both men worked frantically to finally control the spinning, lower the cabin temperature and increase the humidity level to dampen the risk of fire.

Now, to their dismay, and having seen through these life-threatening dramas, they noticed that the craft's guidance system had failed, and even though they could safely initiate their re-entry, they would be well off-target, having climbed to a much higher altitude than planned. They would also have to fire their retrorockets manually. Having oriented the spacecraft for re-entry – no easy task – the return burn went well following guidance from the ground, but the crew knew they now had very little control over where they would land, other than that it would probably be somewhere where Soviet recovery forces might be in the vicinity.

Having successfully negotiated the blistering heat of re-entry, the spacecraft's drogue and main parachutes blossomed, lowering Voskhod-2 to the ground. The crew landed in 2-metre-deep snow within a thick forest situated about 75 kilometres (47 mi.) from the city of Perm in the Ural Mountains, 26 hours and 2 minutes after lifting off from Baikonur.

According to Leonov,

> We landed and opened the hatch. The air was cold, it rushed in. We set up our radio channel and began to broadcast our coded signal. Only after seven hours did a monitoring station in West Germany report that they had heard the coded signal which I had sent.[10]

As the two men waited in their freezing-cold spacecraft, huddled in temperatures of around -25°C (-13°F), they could hear wolves and bears moving through the forest as darkness set in. Under such conditions, they knew they would have to wring any moisture out of their spacesuits to prevent the onset of frostbite. Leonov's suit was filled with so much perspiration from his spacewalk that it was sloshing about in the legs of his spacesuit, right up to his knees. Hours later, they heard the very welcome *thump-thump* of helicopter blades overhead. The pilot and crew were unable to land, but they threw down some warm

Leonov, a talented artist, with a depiction of his history-making spacewalk on 18 March 1965.

clothing – most of which got caught in the treetops – and a bottle of cognac, which unhappily smashed when it hit the ground.

Eventually a recovery crew landed 9 kilometres (5.6 mi.) away and reached the frozen cosmonauts, building a fire and boiling some water for them. The following morning, Leonov and Belyayev donned skis and made their way to a waiting helicopter. From there, they were flown to Perm and then back to the launch site at Baikonur.

On 10 January 1970, five years after commanding the historic flight of Voskhod-2, 44-year-old Colonel Pavel Belyayev died as a result of peritonitis, following complications during an operation on a stomach ulcer. Had the Soviet Union achieved its goal of landing a sole cosmonaut on the Moon, that man would have been the first spacewalker Alexei Leonov. When their lunar-landing programme was scrapped after the success of the lunar-orbiting Apollo 8 mission, Leonov instead became the Soviet commander on the Apollo–Soyuz link-up mission in 1975. Highly regarded as an amiable ambassador for the Soviet–Russian space programme, and a truly talented artist, Major General Alexei Leonov passed away on 11 October 2019, aged 85.

THE FLIGHT OF VOSKHOD-2 would prove to be the last in that programme. Following a dramatic change of Soviet leadership, the focus would eventually shift to a concerted lunar-landing effort, and this was hastened by the unexpected death of Sergei Korolev in the spring of 1966. There were plans in place for another three crewed Voskhod missions, including one flown by two female cosmonauts, Valentina Ponomaryova and Irina Solovyova. These were to be:

- Voskhod-3: a nineteen-day mission to study long-term weightlessness.
- Voskhod-4: a second study of long-term weightlessness with a sole cosmonaut.
- Voskhod-5: a ten-day mission with an all-female crew, initially including the first spacewalk by a woman, to be carried out by Solovyova.

Voskhod-3 came close to being realized, with a launch set for March 1966. Then there were development problems with the parachute system and the ECS (environmental control system) during February and March. A Voskhod booster failure in April caused a further postponement until 25 May, by which time NASA had already launched the first of its two-man Project Gemini missions. The Soviet government, now under new leadership, saw no advantage in allowing yet another Voskhod mission, especially after the death of Korolev earlier that year, and wanted to shift the focus of its space programme towards the Moon. All proposed Voskhod missions were formally cancelled on 1 December 1966.[11] There had also been some discussion and early planning for a further three Voskhod missions (6, 7 and 8) with possible military applications, but these flights were similarly discarded, with a renewed emphasis now on the Soyuz and lunar-landing programmes.

Soviet Voskhod Missions

Flight	Crew	Launch	Landed	Orbits
Voskhod-1	Col. Vladimir M. Komarov Dr Boris B. Yegorov Konstantin P. Feoktistov	12.10.1964	13.10.1964	16
Voskhod-2	Col. Pavel I. Belyayev Lt Col. Alexei A. Leonov	18.03.1965	19.03.1965	17

5

THE TROUBLE WITH SOYUZ

Following the historic Voskhod-2 mission in March 1965, during which Alexei Leonov became the first person – albeit perilously – to complete a spacewalk, there was a lengthy hiatus in Soviet space missions. Meanwhile, NASA began operating the first flights in its two-man Gemini programme. It was an amazingly successful period of activity for the American space agency; over a period of twenty months, NASA managed to launch ten crewed Gemini missions at an average rate of one every two months. Still, there was a confusing silence behind the political boundary between the Soviet Union and the West known as the Iron Curtain. With all the major objectives of the Gemini programme satisfactorily completed, NASA now turned its focus to Project Apollo – the thrust to land astronauts on the Moon by the end of the decade.

There were two principal reasons behind the delay in the Soviet space programme: the development of the next-generation Soyuz (Union) spacecraft was running well behind schedule, combined with the premature death of Chief Designer Sergei Korolev in January 1966 at the age of 55. Despite these setbacks, unrelenting pressure was being applied on the various and now badly disorganized design bureaus to create a new space spectacular, as anxious Soviet leaders could see their previous propaganda advantage being rapidly eroded in the wake of America's outstanding achievements with Project Gemini. Soviet leader Leonid Brezhnev was desperate to show the world that the Moon race was far from over, and like his predecessor, Nikita Khrushchev, began pushing hard for flights to resume, targeting 1967 to coincide with the fiftieth anniversary of the Bolshevik Revolution and the tenth anniversary of the first Sputnik satellite. As a result, haste soon began to overtake caution and systemic shortcutting resulted as the state-driven development of the Soyuz spacecraft continued at an accelerated pace. Meanwhile, with Project Gemini at an end, NASA pressed ahead with its plans for a crewed lunar landing before the end of the decade.

ON 27 JANUARY 1967, despite identifying numerous problems with their Apollo spacecraft and its systems, three NASA astronauts were in final training for an Earth-orbiting test flight scheduled for launch on 21 February. Veteran Mercury and Gemini astronaut Gus Grissom was the mission commander, along with America's first spacewalker, Edward White II, and rookie astronaut Roger Chaffee. On that day, they were involved in what was known as a 'plugs out' flight readiness and countdown demonstration test at the Kennedy Space Center's Launch Complex Pad LC-34. Their spacecraft cabin was internally pressurized using 100 per cent oxygen, but this was not considered a particularly dangerous test of the spacecraft and its systems, as their Saturn IB carrier rocket was not loaded with fuel. Just after 6:30 p.m., a brief electrical arc is believed to have suddenly flared in some faulty wiring. Within moments, the three astronauts were fighting for their lives, trying desperately to open the cabin's pressurized hatch as a ferocious fire fed by pure oxygen began sweeping up the walls, consuming

This metre-long Soyuz model, now on display in Moscow's Polytechnic Museum, was crafted from steel and plastic in Sergei Korolev's OKB-1 design bureau.

and melting everything in its path. Despite their heroic efforts, the astronauts were unable to escape the inferno. Within fifteen seconds of first raising the fire alarm, all three men had perished from inhaling the suffocating toxic fumes. Once it was deemed safe to do so, the distraught pad crews were able to open the heavy main hatch, finding the bodies of the three dead astronauts in the gutted interior.

EVEN AS THE UNITED STATES was mourning the loss of the three Apollo astronauts, the Soviets were moving ahead with an audacious plan to launch not one but two crewed new-generation Soyuz spacecraft. Even the name Soyuz gave a hint as to its forthcoming purpose. The plan involved launching the first Soyuz with a sole cosmonaut on board, to be followed into orbit the next day by Soyuz-2, this time carrying three cosmonauts, two of whom would transfer from their Soyuz craft to the other after docking. As there was no means of transferring directly from one spacecraft to the other, the two men would have to spacewalk across to Soyuz-1. The Soyuz-2 pilot would return alone at mission's end, while the trio of cosmonauts would land in Soyuz-1.

Soyuz-1 pilot Vladimir Komarov.

Komarov with Yuri Gagarin, later revealed as his Soyuz-1 back-up pilot.

The person selected in November 1966 to fly the Soyuz-1 mission was the highly regarded Vladimir Komarov, destined to become the first Soviet cosmonaut to fly a second mission following his earlier assignment as commander of the first Voskhod flight. It would later be revealed that his back-up pilot was Yuri Gagarin, who had fought a tenacious battle to return to space after being politically shelved, as he had become a nationally important person and the Soviet hierarchy did not want to risk him on a second mission. Nevertheless, he had finally convinced the space chiefs to restore his active status, and permission to serve as Komarov's back-up had been given, albeit with reluctance.[1]

Each spacecraft was planned to conduct a four-day mission, with both crews scheduled to demonstrate orbital manoeuvre and rendezvous techniques, and practice an automatic spacecraft docking. The crew assigned to Soyuz-2 comprised mission commander Valery Bykovsky, with another Vostok veteran, Andrian Nikolayev, to serve as his back-up. The other two seats would be occupied by space rookies Yevgeny Khrunov and Alexei Yeliseyev, who would perform the transfer EVA. At a meeting held in Moscow on 25 March, Dr Vasily Mishin, who had taken over as chief designer of the Soviets' lunar programme following the unexpected death of Korolev, announced that the anticipated launch date for Soyuz-1 was 21 or 22 April, while Soyuz-2 would be launched the following day.[2]

On 23 April 1967 the Soviet Union thundered back into space with the successful launch of Soyuz-1, carrying Colonel Komarov. There was a sense of excitement for anyone sleuthing the Soviet space programme for clues on the mission plans. A strong clue was provided when Soviet news agency TASS announced the name of the spacecraft type as Soyuz. This may, of course, have simply been a reference to the USSR, or Union of Soviet Socialist Republics, but many people were convinced that as the name indicated a union it could also herald a planned docking with a second ship. In the Vostok and Voskhod programmes, the first ships to be launched had not carried a numerical designation. Gagarin's spacecraft had been called Vostok, while Gherman Titov's was Vostok-2. The same applied for Voskhod and Voskhod-2. Observers felt that the numeric designation Soyuz-1 had significant implications for a possible rendezvous mission, and they were right: Soyuz-2 was already being prepared for lift-off from the same launch pad.

Just as Komarov achieved orbit, things began to go seriously wrong aboard Soyuz-1 and his immediate concern became trying to resolve several issues. First, one of the two solar panels that would supply power to the spacecraft failed to deploy once orbit was attained. The automatic orientation system also failed, and despite his best efforts, Komarov was unable to align the spacecraft manually. A few days prior to the flight, a group of leading Soviet engineers connected to the space programme had identified around two hundred existing faults in the spacecraft's systems, including the troublesome orientation system, but their pleas to delay the flight until those technical problems could be resolved fell on deaf ears within the Kremlin.

The situation in space was now becoming so serious that a post-ponement of Soyuz-2, due to launch the next day, was considered but initially rejected. It was felt that ground control, working with Komarov, could overcome these technical issues. It was a reckless decision brought about by the pressure from the corridors of power to achieve this latest space spectacular.

As the hours passed, it became increasingly evident that Komarov would not be able to rectify the orientation problem. In desperation, he had even tried kicking the cabin wall near the outside solar panel,

The crew of the aborted Soyuz-2 mission: Yevgeny Khrunov, Valery Bykovsky and Alexei Yeliseyev.

but it would not deploy. Eventually, after 27 hours of frantic efforts by an exhausted Komarov, a decision was reached – with much reluctance – to postpone the launch of Soyuz-2 and stand the crew down until further flight plans could be promulgated. Komarov was then instructed to prepare the Soyuz descent module for a manual re-entry, with which he was largely unfamiliar. The first attempt failed when the spacecraft could not be properly aligned for re-entry, but a second, on the nineteenth orbit, was successful. Much to the relief of Komarov and ground control, Soyuz-1 finally began plunging back into the atmosphere.

Komarov's re-entry went as planned, and he continued to record his reactions as the Soyuz craft plummeted down to the designated landing area in northern Kazakhstan. Just as he began to anticipate his landing, things went disastrously wrong. The small drogue chute at the top of the spacecraft deployed as scheduled, but when it was time for the main parachute to be released and unfurl, nothing happened. An overlooked fault in the chute container during production meant that the lining of the container was quite rough, and this prevented the main parachute from deploying. The reserve chute was then meant to deploy automatically, but it became tangled in the lines of the drogue chute owing to the disrupted airflow and would not open.

Komarov's widow, Valentina, at his funeral in Moscow's Red Square, 26 April 1967.

In the last few seconds of his life, Komarov would have realized that his fall was not being slowed down, and he would have known that he was about to die. The Soyuz spacecraft smashed into the ground near Karabutak in western Kazakhstan at more than 140 kmph (87 mph), killing him instantly. Small braking rockets, meant to fire moments before landing, exploded on impact, and the shattered spacecraft erupted into flames. When the recovery crews finally arrived on the scene, they threw buckets of soil onto the twisted remains of the spacecraft, trying to extinguish the flames, but they could do little else. On hearing this distressing news, General Nikolai Kamanin ordered that the debris be cleared away and Komarov's body retrieved. As one of those rescuers later recalled,

> The group's physicians set to work – they shovelled away the top layer of dirt from the top of the mound from the hatch cover. After the dirt and certain parts of instruments and equipment were removed, the cosmonaut's body was found lying in the centre chair. The physicians cleaned the dirt and the remnants of the burnt helmet phone from the head. They pronounced the death to be from multiple injuries to the cranium, spinal cord and bones.

They could do little else but extract Komarov's charred remains for transportation back to Moscow.[3]

The death of Vladimir Komarov was a tremendous blow to the Soviet space programme, and the entire nation fell into a shocked, collective sadness. His remains were inurned in the Kremlin Wall with full military honours on 26 April. The usual shroud of secrecy surrounded the flight of Soyuz-1, which brought forth some ugly, unjust rumours about the circumstances of Komarov's death. One concerned his wife, Valentina, being collected and taken to mission control to bid an emotional farewell to her doomed husband when it appeared he could not be brought back to Earth alive. Prime Minister Alexei Kosygin was also said to have talked with Komarov, telling the cosmonaut that his country was proud of him and his forthcoming sacrifice for the space programme. During the descent module's re-entry, American listening stations were reported to have eavesdropped on Komarov's pitiful cries as he plummeted towards the ground, cursing and renouncing a government that had ordered him to carry out a flight in a trouble-plagued craft launched well before it was ready.

These disrespectful falsehoods gained traction for a while, until they were proved false. First, the programme's mission control was situated at Yevpatoriya in the Crimea, while Valentina Komarov would have been at home in Moscow. Additionally, mission control would have lost contact with the cosmonaut shortly after the descent module separated from the other two spacecraft modules and began its re-entry. This was a perfectly normal occurrence for any returning spacecraft: the ionized air surrounding a vehicle returning through the ferocious heat of re-entry causes a radio blackout lasting several minutes.

In 1992 the author corresponded with unflown cosmonaut Aleksander Petrushenko, who was the last person in contact with the doomed cosmonaut. During their exchange, Komarov calmly reported to Petrushenko that he had carried out the final orientation manoeuvre and was preparing for his return to Earth.[4]

If there was an upside to Komarov's loss, it was that his early return from space undoubtedly saved the lives of the three cosmonauts preparing to fly into space aboard Soyuz-2. The two scheduled to transfer to Soyuz-1 in orbit would have perished with him, while subsequent investigations revealed that Soyuz-2's parachute container carried exactly the same design fault as that of Soyuz-1, which would in all likelihood have resulted in the death of the third cosmonaut, Bykovsky.

A month after the loss of Vladimir Komarov, distressed fellow cosmonauts Yuri Gagarin and Alexei Leonov criticized Vasily Mishin's 'poor knowledge of the Soyuz spacecraft and the details of its operation; his lack of cooperation in working with the cosmonauts in flight and training activities', and asked the head of cosmonaut training, Nikolai Kamanin, to cite Mishin in the official report into the loss of Soyuz-1. Leonov went so far as to describe Mishin as 'hesitant, uninspiring, poor at making decisions, over-reluctant to take risks and bad at managing the cosmonaut corps'.[5]

Both the United States and the Soviet Union had suffered tragic crew losses in 1967, and both nations would subsequently undergo a considerable period of reflection and rebuilding before they could resume the monumental challenge of sending the first landing crew to the Moon.

6

LOSING THE MOON

When Vladimir Komarov's parachutes failed to deploy and he was killed following the otherwise survivable re-entry of his Soyuz-1 spacecraft, the Soyuz programme became mired in deep trouble, and all future crews were forced to stand down while the cause was being investigated. An inquiry found that not only was there an excess of haste involved in trying to launch the spacecraft before it was fully tested, but Komarov had shown incredible courage and skill in mostly overcoming the attitude problems that cut his mission short, allowing him to return to Earth. The loss of Soyuz 1 was found to be entirely due to a flawed and unchecked parachute system.

The destruction of Soyuz-1, along with a decorated cosmonaut, served to highlight a signature weakness in the Soviet space programme – a lack of substantial funding. Following the death of Korolev in January 1966, just two days after turning 59, his successor, General Vasily Mishin, became the head of OKB-1 (later redesignated TSKBEM). While Korolev's numerous successes meant he could request and secure state funding for his space projects, Mishin did not enjoy anywhere near the same level of trust or confidence. Mishin would always struggle to obtain financial backing, which led to the programme losing much of its momentum, suffering unsustainably severe cutbacks and spawning shortcuts in the monumental task of reaching the Moon ahead of the Americans.

As work proceeded on rectifying identified faults within the spacecraft, which would allow the programme to resume, the Soviet Union suffered another unexpected and paralysing blow on 27 March 1968

First spaceman Yuri Gagarin in the cockpit of an MiG jet, 1968.

when Yuri Gagarin was killed during a training flight in a MiG-15 jet fighter.

Following the loss of his friend Komarov, Gagarin had once again found himself grounded, regarded as being too heroic a person to be lost in another possible space disaster. To his dismay, he was handed a mundane desk job. He never gave up, however, and his first step to regaining his active status was getting back to flying. His management superiors reluctantly agreed, and Gagarin's stand-down order was rescinded on 13 March 1968, meaning he could now operate a MiG fighter, but only under the watchful eyes of an accompanying instructor.

Two weeks later, on 27 March, Gagarin strapped himself into a two-seat MiG-15 trainer along with his supervising instructor, Colonel Vladimir Seryogin. After completing the necessary pre-flight checks, they took off from the Chkalovsky Air Force Base, situated near the cosmonaut training centre, at 10:19 a.m. Once they were established in flight, they requested a change of course, which was standard practice and granted. It would be the last communication heard from the aircraft, which suddenly vanished from the screens of the ground controllers.

Deeply concerned, the controllers requested that search planes be immediately despatched. Soon, everyone's worst fears were realized when wreckage of an aircraft was spotted in a birch wood near the village of Novoselova, some 89 kilometres (55 mi.) east of Moscow.

While the actual cause of the crash has never been established, the official report suggests that the most likely explanation is that the MIG had gone into a violent spin after flying into the trailing vortex of an unidentified fighter jet and crashed before the two pilots could recover the fighter, level out and safely regain altitude.[1]

Yuri Gagarin had turned 34 just three weeks before he died. His remains were inurned in the Kremlin Wall, amid an outpouring of grief by a shocked nation who remembered a smiling young cosmonaut who had achieved everlasting fame as the first person to fly into space.

THE SOVIET SOYUZ programme finally resumed with the launch of the Soyuz-2 spacecraft on 25 October 1968, ending an eighteen-month hiatus. During this time, the faulty parachute system had been totally overhauled and the integrity of the Soyuz craft verified. In addition, an automatic test docking in orbit had been achieved by two unmanned craft identified only as Kosmos-186 and Kosmos-188 on 30 October 1967.

On this occasion, the Soyuz-2 vehicle was unmanned, but as with the originally planned link-up between Soyuz-1 and -2, the revised

Stunned Muscovites holding banner photographs of Vladimir Seryogin and Yuri Gagarin during the Red Square funeral service.

Soyuz-3 pilot Georgi
Beregovoi.

programme called for Soyuz-3 to be launched the following day, with
47-year-old Lieutenant General Georgi Beregovoi in command.
Beregovoi, a military test pilot with the Soviet Air Force, was selected
as a cosmonaut in January 1964, and his first assignments were for the
Vostok-10 and Voskhod-3 missions, which were both later cancelled.

Earlier there had been a tentative push for Gherman Titov to be
awarded a second space flight despite his still-unexplained illness on the
Vostok-2 mission. Titov had remained in mission training, but in the
eyes of many members of the Politburo (the principal policy-making
committee in the Soviet Union), he had become something of a loose
cannon following his flight: despite severe warnings about his attitude,
he had been freely womanizing and drinking and was fond of driving
too fast. For the time being, his maverick behaviour was being kept
under wraps, but questions would undoubtedly be asked as to why the
popular cosmonaut had not been given a second flight. Although he
could have been assigned to Soyuz-3, it was felt that this was too critical
a mission for Titov, and an experienced piloting hand was needed.
Nikolai Kamanin, the head of cosmonaut training, strongly endorsed
Beregovoi, who had served under him during the Second World War
and flown 185 combat missions. In the 1950s he became a proficient
test pilot, flying zero-launch rocket-boosted fighters. Kamanin's rec-
ommendation held a lot of weight, and Beregovoi was awarded the

Soyuz-3 assignment, with Vladimir Shatalov as his back-up pilot. Titov found himself returned to standby duties for a possible later mission, which would never materialize.

Beregovoi's main objectives on the Soyuz-3 mission were to perform comprehensive testing of the spacecraft's systems and to achieve a hard docking with a passive target vehicle – the unmanned Soyuz-2. His flight began on 26 October 1968, a day after Soyuz-2 had been launched and placed into orbit and eighteen months after the loss of Vladimir Komarov aboard Soyuz-1.

Once Beregovoi had achieved orbit, the Igla (Needle) automatic rendezvous system locked onto the Soyuz-2 spacecraft and brought him to within about 180 metres (590 ft) of the target vehicle. As scheduled, Beregovoi then took over manual control of his spacecraft and closed the gap between them. At this point, his problems began as he continually failed to dock with the passive Soyuz-2. While he was able to close the gap between the two craft to just 1 metre (3 ft), his three attempts at docking all failed, during the course of which he expended almost all of his orientation fuel. He was finally instructed to abandon the docking.

A frustrated Beregovoi would spend a further three days in space. He did manage to conduct some complex testing of the spacecraft's systems, but, essentially, his flight would be deemed a failure. Meanwhile, contradictory information supplied to the world's press suggested that all of the mission's major objectives had been met, implying that there had only been plans for a rendezvous with Soyuz-2, not a docking. Caution, the report stated, was being employed after the loss of Soyuz-1 the previous year. Beregovoi's flight ended with a successful touchdown less than 10 kilometres (6 mi.) from the target zone after 3 days, 22 hours and 48 minutes in space. His performance on the mission was reviewed as poor and he would never receive a second flight. Eventually, he took up a position at the cosmonaut training centre, becoming director of the facility in 1972, a position he held until 1986.[2]

In an interview given to Russia's *Trud* newspaper in April 2002, Voskhod cosmonaut and OKB design engineer Konstantin Feoktistov stated that Beregovoi had

> committed the grossest error . . . He did not turn his attention
> to the fact that the ship to which he was meant to dock was

[upside down in relation to his own spacecraft]. The flashing lights of the unmanned ship proved to be on top, and should have been below. Therefore, the approach of Soyuz-3 [caused] the pilotless object to turn away. In these erroneous manoeuvres, Beregovoi consumed all the fuel intended for the ship docking.[3]

In January 1969 two crewed missions were launched. On 14 January Soyuz-4 achieved orbit; the sole occupant of the spacecraft was Vladimir Shatalov on his first mission. The following day, the Soyuz-5 mission began, now with three cosmonauts on board: Boris Volynov (commander), Alexei Yeliseyev and Yevgeny Khrunov. Like Shatalov, all three men were on their first flight into space. On the 34th orbit of Soyuz-4, Shatalov took over manual control of his spacecraft and carried out an incident-free docking with Soyuz-5. A whole new space 'first' would begin with an in-orbit crew transfer. As there was no transfer tunnel between the vehicles, an exterior EVA was required.

Once Yeliseyev and Khrunov had donned their Yastreb spacesuits in the orbital module of Soyuz-5, they were thoroughly checked by Volynov, who then made his way back into the descent module, sealed the hatch and depressurized the orbital module. The two spacewalkers then opened their hatch and moved outside, making their way hand-over-hand to the

The crewmembers of Soyuz-4 and Soyuz-5 relaxing with a group of instructors during a training exercise. At front, from left: Yevgeny Khrunov, Alexei Yeliseyev, Boris Volynov and Vladimir Shatalov.

Flight Engineer Valery Kubasov and Pilot Commander Georgi Shonin: the crew of Soyuz-6.

open hatch of Soyuz-4, where Shatalov was waiting to greet them. Once they were safely inside, the hatch was sealed and the spacecraft repressurized. The two Soyuz craft then undocked and their separate flights continued. Soyuz-4, now with a crew of three, touched down on 17 January. Volynov endured a bumpy landing the day after.

Nine months later, on 11 October, by which time America had successfully placed the landing crew of Apollo 11 onto the Moon's surface, another Soviet space spectacular began to unfold. On that day, Soyuz-6 was launched from the Baikonur pad, carrying Georgi Shonin and Valery Kubasov into orbit.

A day later, a further three cosmonauts were launched into orbit aboard Soyuz-7: Anatoly Filipchenko (commander), Vladislav Volkov and Viktor Gorbatko. Twenty-four hours on, Soyuz-8 left the launch pad: for the first time, three crewed spacecraft were in orbit simultaneously. The third crew comprised Vladimir Shatalov and Alexei Yeliseyev, both of whom had flown on the previous tandem mission. Much to the amazement of Western observers, the three ships did not link up in orbit. In fact, Soyuz-7 was meant to dock with Soyuz-8 while observed by the two cosmonauts aboard Soyuz-6 some 50 metres (160 ft) distant, during which activity one crewmember from each craft would transfer to the other linked ship. However, the Igla automated docking system failed despite several attempts. Ground control would

The three Soyuz-7 cosmonauts (from left): Vladislav Volkov, Viktor Gorbatko and Anatoli Filipchenko.

not permit the crews to try an unscheduled manual docking, and it was abandoned. All three crews later landed safely.

Launched on 1 June 1970, Soyuz-9 was piloted by commander Andrian Nikolayev on his second space flight, accompanied by Vitaly Sevastyanov. This would be a flight endurance test in preparation for extended stays aboard future space stations. Scientific, technical, medical and biological studies would be carried out throughout their lengthy space flight. The two cosmonauts would return to a safe landing after a then-record eighteen days in orbit.

BUT WHAT OF SOVIET PLANS to place cosmonauts on the Moon? It was in 1967 that Soviet academician Georgi Arbatov told Western journalists, 'The very next milestone in our conquest of space will be a manned circumnavigation of the Moon, and then a lunar landing.'[4]

During budget hearings, NASA administrator James Webb informed Congress that KH-8 spy satellite photographs had identified a massive rocket poised upright on a Baikonur launch pad in Kazakhstan. Webb stated that a rocket of this size could potentially be twice as powerful as America's Saturn v, currently under development, and was obviously intended to send cosmonauts to the Moon. Members of Congress were not permitted to see these top-secret photographs and many decided that this was nothing more than a political ploy by Webb to obtain additional funding for NASA. That may have seemed the case, but Webb's information would prove to be correct; the giant Moon rocket, later identified as N-1, was a reality.

A year before his tragic death aboard Soyuz-1, Vladimir Komarov had declared, 'I can positively state that the Soviet Union will not be beaten by the United States in the race for a human being to go to the Moon. The u.s. has a timetable of "1969 plus X", but our timetable is "1969 plus X minus one"!' – meaning that the Soviet Union intended to get to the Moon ahead of NASA's astronauts.[5]

Vladimir Shatalov and Alexei Yeliseyev: the crew of Soyuz-8.

Komarov had good reason to make this statement, as the Soviet space programme, once masterminded by Sergei Korolev, had been rapidly developing a means of transporting cosmonaut crews to and from the Moon, in preparation for the ultimate goal of a manned landing. Both programmes – the Earth- and lunar-orbiting flights utilizing Proton rockets and the more complex lunar-landing missions launched atop the far more powerful N-1 booster – were under development at the same time. Had these programmes evolved as planned, and had Korolev not been succeeded by Vasily Mishin, it is highly likely that Alexei Leonov would have been the first person to plant that first footprint on the Moon's dusty surface. As Leonov wrote in 2004:

> I was undergoing intensive training for a lunar mission by this time. In order to focus attention and resources our cosmonaut corps had been divided into two groups. One group, which included Yuri Gagarin and Vladimir Komarov, was training to fly our latest spacecraft – Soyuz – in Earth orbit. (Korolev had begun theoretical work on Soyuz as early as the 1950s, and construction of the spacecraft, a modified version of which still flies today, had begun several years before his death.)
>
> The second group, of which I was commander, was training for circumlunar missions in a modified version of Soyuz known as the L-1 [alternatively LK-3], or Zond, and also for

An unnamed artist's impression of the orbital link-up and crew transfer between Soyuz-7 and Soyuz-8, while the nearby crew of Soyuz-6 observe the operation.

Andrian Nikolayev and Vitaly Sevastyanov set a new space endurance record aboard Soyuz-9.

lunar-landing missions in another modified Soyuz known as the L-3 [or LK-3]. Vasily Mishin's cautious plan called for three circumlunar missions to be carried out with three different two-man crews, one of which would then be chosen to make the first lunar landing.

The initial plan was for me to command the first circum-lunar mission, together with Oleg Makarov, in June or July 1967. We then expected to be able to accomplish the first Moon landing – ahead of the Americans – in September 1968.[6]

The first group of (Earth and lunar orbital) cosmonauts, as mentioned by Leonov and under the command of Vladimir Komarov – who could elect to fly – also comprised cosmonauts Yuri Gagarin, Andrian Nikolayev, Valery Bykovsky, Yevgeny Khrunov, Viktor Gorbatko, Georgi Grechko, Vitaly Sevastyanov, Valery Kubasov and Vladislav Volkov. The second (landing) group was made up of Leonov, Pavel Popovich, Pavel Belyayev, Boris Volynov, Pyotr Klimuk, Oleg Makarov, Anatoly Voronov, Nikolai Rukavishnikov and Yuri Artyukhin.

In August 1967 Russia's TASS news agency reported that ten cosmo-nauts, including Alexei Leonov, were practising recovery techniques

Under different circumstances, cosmonaut Alexei Leonov may have been the first person to land and walk on the Moon.

normally associated with water splashdowns. This confused Western space analysts, as all previous Soviet space flights had ended with a parachute touchdown on land. There was no real secret behind this: the cosmonauts were simply undergoing contingency training in the event of an off-course landing in the sea. This would happen a year later when the unmanned Zond-5, launched on 15 September 1968, became the first spacecraft to successfully circle the Moon and return to Earth, in this case with an unplanned splashdown in the Indian Ocean six days later. It had been planned to land the craft in Kazakhstan, but a guidance problem forced the returning module into a steeper, high-speed ballistic re-entry, ending with a night splashdown in the sea 105 kilometres (65 mi.) from a Soviet ship of the Search and Rescue Service. Fortunately, the spacecraft was airtight and remained buoyant until it could be recovered.

Zond-5 was one of a series of stripped-down Soyuz spacecraft without the forward orbital module. A biological payload was on board, comprising two tortoises, wine flies, worms, plants, seeds, bacteria and other living matter. They became the first lifeforms to travel around the Moon and return safely. During the mission (and much to the confusion of Western monitors), taped voice conversations were even beamed back to Earth as a communications test. Significantly, one Zond even underwent a test of a launch escape system. Unmanned craft require no such escape system.

It would be revealed some years later that the top-secret Zond-5 mission was originally planned to fly two cosmonauts around the Moon. Earlier component failures within the Proton-K rockets on the Zond 1968A and Zond 1968B missions resulted in the destruction of both spacecraft in launch explosions, leading the Soviets to launch an unmanned test mission instead, fearful of the negative publicity that would follow a possible loss of a lunar crew.[7] America, it seems, came very close to coming second in the race to the Moon.

A month after the successful end of the Zond-5 mission, cosmonaut Gherman Titov was in Acapulco to observe Olympic yachting trials, and during one interview he bragged, quite unwisely as it transpired, that he would be the pilot on the first circumlunar mission. A week later, he was forced to retract that claim over Radio Moscow. Then, on 10 November, Zond-6 was launched on a mission to loop around the Moon, carrying photography equipment and another biological payload including bacteria, flies and tortoises.

As it was seen as a precursor to a possible manned circumlunar flight – perhaps as early as December 1968 – several key figures attended the launch of Zond-6, including the spacecraft's chief designer, Vasily Mishin, and the head of cosmonaut training, Nikolai Kamanin, together with the cosmonauts then in training for upcoming lunar missions.

There was understandable jubilation when the Proton-K rocket launch went well, and four days later, Zond-6 flew around the Moon, coming within 2,420 kilometres (1,504 mi.) at its closest approach. Several colour and black-and-white photographs were taken of the lunar surface. Then, as the spacecraft headed home, it unexpectedly depressurized, destroying all the biological specimens on board. To compound this mishap, the landing parachutes failed and the spacecraft smashed into the ground some 70 kilometres (43 mi.) northeast of the launch site. One item salvaged from the wreckage was the photograph packs, secured inside fortified metal cassettes. Some of the film was torn and overexposed, but the surviving images were printed and released to the world, accompanied by a statement suggesting that the flight had been a complete success.

Despite this setback, veteran cosmonaut Georgi Grechko was adamant that Soviet cosmonauts could have easily sent a crew on this flight to orbit the Moon ahead of the Americans, but the space chiefs failed to get any support for doing so from an over-cautious Kremlin Politburo, who quashed the idea. 'The only thing [needed] was to get the final "go-ahead,"' Grechko stated, 'and we would have been around the Moon in an instant!'[8]

The problems with Zond-6 and other failures caused the Soviet Union to delay any ambitious plans for a crewed flight to the Moon, and when the December 1968 mission of Apollo 8 changed from being an Earth-orbit test of the lunar-landing module to a lunar-orbiting mission, it signalled the end of any further attempts to send cosmonauts to the Moon. With a crewed Apollo landing now projected for mid-1969, Soviet chiefs stoically continued their denials that they had

A full-size mock-up of an N-1 rocket on display at the Baikonur launch site.

ever planned to send cosmonauts on a lunar journey. The chairman of the Soviet Academy of Sciences, Mstislav Keldysh, even told Swedish news reporters in October 1969 that the Soviet Union had abandoned plans for manned flights to the Moon, instead devoting its efforts to placing space laboratories into orbit around the planet. 'We no longer have any time plan for manned Moon trips,' he declared. 'Right now we are concentrating on building big satellite space stations.'[9]

By far the biggest setback to Soviet plans for lunar conquest was the failure of its massive N-1 rocket, originally designed by Korolev. This was meant to be the equivalent of America's Saturn V, which would carry astronauts on their journeys to the Moon. By way of comparison, the N-1 stood almost as tall as the Saturn V at 105 metres (344 ft). As developed by the Kuznetsov design bureau, the N-1 was capable of generating 4,620 tonnes of thrust, the highest at lift-off of any rocket ever developed, eclipsing even the 3,400 tonnes of thrust generated by America's Saturn V and its five massive F-1 engines. The first (or booster) stage of the N-1 utilized a total of thirty NK-15 engines arranged in two concentric circles at the foot of the booster, with 24 mounted around an outer perimeter ring and another six on a second, inner ring. Each NK-15 engine could provide a maximum 154 tonnes of thrust at lift-off. This meant that all thirty, firing simultaneously, were needed just to get the brutally heavy rocket off the launch pad. However, the second and third stages were far less efficient, greatly reducing the payload capacity. Overall, the N-1 programme had become insanely complex and doomed to failure due to gross underfunding and a loss of focus, given the size and complexity of the project.

Whereas the Saturn V would transport three astronauts on their lunar flight, weight limitations imposed on the N-1 meant it could only carry a crew of two cosmonauts. During the landing phase, only one cosmonaut would make the final descent to the lunar surface.

In addition, the design and sophistication of the Apollo spacecraft far exceeded that of the Soviets. The command module boasted a transfer tunnel through to the attached lunar module, making internal movement between the two craft a simple task. No such facility existed between the Soyuz spacecraft and the LK-3 lunar module. This meant that the landing cosmonaut – expected to be Alexei Leonov – would have to open the Soyuz hatch and spacewalk across to the LK-3. He would then open its hatch, move in to a standing position, close the hatch and fly the LK-3 solo down to the surface of the Moon. All of

An engineering model of the LK-3 lunar lander on temporary display at London's Science Museum.

this involved potentially greater risk than that faced by the Apollo crew. Once he was ready to leave the Moon, the cosmonaut would fly his craft back into lunar orbit, complete a docking with the aid of the second orbiting cosmonaut and then transfer back into the Soyuz with a second spacewalk. It would be an exhausting and hazardous procedure for both crewmembers.

This scenario would remain hypothetical, however, as there were massive problems inherent in the N-1 launch vehicle. The thirty small engines clustered around the central first stage of the rocket had to fire in complete synchronization, and just one mishap in this precise choreography could result in cataclysmic failure. Whereas the mighty F-1 engines that powered NASA's Saturn V rocket were thoroughly tested and certified at purpose-built testing grounds, no such facility was made available to test the N-1's engines. It remained a recipe for disaster, and so it would prove.

The first attempt to launch an N-1 rocket (No. 3L) took place on 21 February 1969, carrying an unmanned L1 spacecraft payload. Sixty-eight seconds into the launch, the flight came to a premature end when a leaking oxidizer pipe caused a fire to erupt at the base of the first stage. While the L1's escape system activated, pulling the unmanned craft away from the booster and landing safely by parachute, the range safety officers had no option but to initiate the destruction of the wayward rocket.

The second launch attempt (N-1 No. 5L) took place on 3 July 1969, some six weeks after Apollo 10 had carried out a rehearsal flight ahead of the planned Apollo 11 landing mission. This time, the N-1 managed to climb to 100 metres (328 ft) but then started to disintegrate and veered wildly off course before crashing back onto one of the cosmodrome's two launch pads.

There was a massive explosion and a huge black cloud belched into the sky. The pad suffered substantial damage, but fortunately there were no casualties, apart from any lingering ambitions to beat the United States to the Moon. Over the next five years, there would be two further attempts to launch N-1 rockets, both ending in failure. By now, the Kremlin chiefs were afraid of taking the responsibility should another cosmonaut die on a space mission. The N-1 programme was finally cancelled by the Soviet government in June 1974. Alexei Leonov described the cancellation as a 'devastating personal blow . . . I argued very hard that we should continue with our work, but the higher powers were adamant. The lunar groups which I had commanded and trained with for three long years were disbanded.'[10]

There is one overriding conclusion that can be drawn from the Soviet Union's failure to reach the Moon ahead of the Americans: the loss of Vladimir Komarov aboard the Soyuz-1 spacecraft and the complete debacle of the N-1 rocket programme both happened after the death of Sergei Korolev. Up until then, he had overseen an uninter-rupted series of successful space missions dating back to the post-war biological rocket tests flights, many involving canine subjects, the launch of Sputnik in 1957 and the orbital flight of Yuri Gagarin in 1961. With his death, the Soviet space programme rapidly became rudderless and fell into a rapid decline. While he was alive, Korolev, a brilliant manager, was able to ward off the petty meddling and ambitions of competitive designers such as Mikhail Yangel, Vladimir Chelomei, Vasily Mishin and Valentin Glushko, but his untimely loss would quickly prove catastrophic to Soviet space plans. His importance to the early decades of the Soviet space programme can never be underestimated, and should never be forgotten.

7

A TRAGIC SETBACK

Following the continuing success of America's Apollo programme, and with the winding down of their own lunar-landing ambitions, the Soviet Union began turning its attention instead to long-duration space flights aboard orbiting space laboratories. The first of these operational space stations was named Salyut-1, the Russian word for 'salute', in a tribute to the memory of the world's first space traveller, Yuri Gagarin.

Salyut-1 was launched atop a Proton-k carrier rocket from the Baikonur Cosmodrome on 19 April 1971. Several minutes later, it had settled into a planned orbit with an apogee (highest point) of 222 kilometres (138 mi.) and a perigee (lowest point) of 200 kilometres (124 mi.). The station, constructed with only a single docking port, comprised five basic segments: a transfer compartment, a main compartment, two auxiliary compartments and the Orion-1 Space Observatory. In all, it measured 14.6 metres long (48 ft) and at its widest was 4.25 metres (14 ft) in diameter. It was not an original design but a man-rated version adapted from a top-secret, habitable military reconnaissance station called Almaz then under development by Vladimir Chelomei and his team at the OKB-52 design bureau.[1]

Under orders issued from the Kremlin in February 1970, the Almaz hardware and associated plans from Chelomei's bureau were transferred across to the TSKBEM bureau, with expectations that a Soviet space laboratory could and should be launched ahead of NASA's planned Skylab station. Among many design changes, Korolev's bureau was tasked with converting the military station to an ostensibly civilian research facility.

Unknown artist's impression of a Soyuz spacecraft preparing to dock with the Salyut-1 space station.

They would also create a docking bay suitable for Soyuz spacecraft, in lieu of the previously planned TKS special transport spacecraft, developed to carry cosmonauts and supplies to the Almaz station.

With the announcement of the launch of Salyut-1 came the expectation that it would soon be receiving its first visitors, and that proved to be the case. Just four days later, aboard Soyuz-10 a crew of three cosmonauts was launched on an orbital chase of the station, setting the stage for a rendezvous and docking. The TASS news agency reported that the cosmonauts would carry out 'joint experiments' with Salyut-1. Two of the crew, mission commander Colonel Vladimir Shatalov and Alexei Yeliseyev, were on their third (and final) space flights, while Nikolai Rukavishnikov, who would also finish his cosmonaut career with three missions, was experiencing his first flight into space.

Following his flight aboard Soyuz-4 (landing with Yeliseyev on board), Shatalov had stated, with an insider's view,

> I think that our flight opens the way to more sophisticated and varied experiments. Indeed the future lies not with short-duration flights. We, for instance, were always aware of an acute shortage of time for our research. We were eager to learn and see more and bring the findings back to Earth. We were in a hurry. The future of course lies with orbital stations made up of several ships. There is no point in bringing such a station to Earth. Orbit crews may be regularly changed.[2]

An alignment problem caused by an issue with the Igla orientation system resulted in Shatalov taking over manual control of Soyuz-10, and he successfully completed the first docking with an orbiting space station. Unfortunately, Shatalov reported soon afterwards that a light confirming a hard dock with Salyut-1 had not illuminated. This indicated that they had not fully docked to the station, which was confirmed by ground control. Shatalov tried firing their manoeuvring engines to push the two vehicles together, but the problem remained.

Four orbits later, orders came through to fully undock and try again, but this attempt also proved unsuccessful. After a period of deliberation by controllers, a decision was made to abandon any further attempts and bring the crew home. During the re-entry, some toxic fumes began filtering through the interior of the descent module, causing Rukavishnikov to pass out, but he would fully recover. Following their parachute descent over the steppes of Soviet central Asia, the crew then came dangerously close to landing in a wide lake, which could have proved fatal, but they had no control over the descent and could only hope they would miss it. Much to their relief they touched down just 50 metres (164 ft) from the water's edge, a little under two days after they had launched from the Baikonur Cosmodrome, which lay 515 kilometres (320 mi.) to the southwest.

Once the cause of the docking problem had been established, the launch of a second mission to Salyut-1 could proceed, with a newly engineered docking mechanism fitted to the Soyuz-11 spacecraft. As the flight profile was much the same as the previous mission, the fully trained Soyuz-10 back-up crew received the assignment. This crew comprised commander Alexei Leonov together with flight engineer

The crew of Soyuz-10 at the Baikonur cosmodrome prior to their flight. From left, Nikolai Rukavishnikov, commander Vladimir Shatalov and Alexei Yeliseyev.

Valery Kubasov and systems engineer Pyotr Kolodin. The third or reserve crew for the Soyuz-10 mission would now move up and serve as their back-up crew. This crew comprised 43-year-old rookie commander Georgi Dobrovolsky, 36-year-old flight engineer Vladislav Volkov (who had been a member of the Soyuz-7 crew in 1969) and 37-year-old systems engineer Viktor Patsayev.

Then, prior to the launch date, fate would step in and deal a curious hand, as related by Alexei Leonov:

> Shortly before the mission Kubasov developed a problem with his lungs. It turned out later that he was allergic to a chemical insecticide used to spray trees near the cosmodrome in Baikonur. At first a member of the back-up crew, Vladimir [sic] Volkov, was transferred to my crew in Kubasov's place. Then, eleven hours before the launch, the entire crew was changed. It was feared that Kubasov might have a lung infection, and might have transmitted it to Kolodin and myself . . . If Kolodin and I had not already come down with an infection, I didn't see any reason why we should fall ill.[3]

Despite Leonov's angry protestations on behalf of his crew, the decision had been made. Dobrovolsky, Volkov and Patsayev would take on the

The originally assigned crew of the ill-fated Soyuz-11 mission: Valery Kubasov, Alexei Leonov and Pyotr Kolodin.

Prior to launch, the crew of Soyuz-11 perform the ritual walkout to the bus that will carry them to the launch pad. From left: Vladislav Volkov, Georgi Dobrovolsky and Viktor Patsayev.

Soyuz-11 mission. It was a decision that in all likelihood saved Leonov, Kubasov and Kolodin, but cost the members of the replacement crew their lives.

SOYUZ-11 WAS LAUNCHED from the Baikonur Cosmodrome on 6 June 1971, while the Salyut-1 space station was completing its 779th orbit of the planet. In announcing this new flight to the space station, TASS would only reveal that the crew's mission would be longer and more complex than that of Soyuz-10. The news agency gave no indication of how long the three would stay aboard Salyut-1. In a pre-launch news conference, mission commander Georgi Dobrovolsky simply stated that the crew had trained 'for work on the ferry craft, for docking with the Salyut, and for engineering, astro physical and medical experiments'.[4]

Six hours after lift-off, the crew manually corrected the spacecraft's orbit and reported that all systems were performing normally before settling down for a ten-hour rest period. The docking procedure would require all their skill and attention, and they needed to be fresh and alert.

The following day, the Igla automatic docking system guided Soyuz-11 to within 150 metres (492 ft) of Salyut-1, at which point Dobrovolsky took over manual control of the spacecraft and achieved a successful soft docking, following which a mechanical coupling

The crew of Soyuz-11 aboard their spacecraft. From left: Vladislav Volkov, Georgi Dobrovolsky and Viktor Patsayev.

secured a hard dock, with electrical and hydraulic systems then connected. Once checks had been made and the integrity of all components was verified, the hatch was opened and the three cosmonauts moved into Salyut-1, becoming the first crew ever to occupy an orbiting space station. Ahead of them lay a tenure of no more than 25 days.

Over the next three weeks, viewers in the Soviet Union and around the world were able to watch grainy television images of the three cosmonauts as they made daily reports from orbit, emphasizing the scientific, medical and observational work they were carrying out. Viktor Patsayev became the first person to operate a telescope – the Orion 1 Space Observatory – outside Earth's atmosphere. He also became the first person to enjoy a birthday in space, on 19 June.

On their eighteenth day, the crew surpassed the endurance record held by Nikolayev and Sevastyanov on the Soyuz-9 mission, and five days later they began preparations for their return to Earth. As TASS later reported, the three cosmonauts 'transferred the materials of scientific research and the logs' to the descent module of Soyuz-11. 'After completing the transition operation, the cosmonauts took their seats in the Soyuz-11 ship, checked the on-board systems and prepared the ship for unlinking from the Salyut station. The crew reported to Earth the unlinking operation passed without a hitch and all the systems were functioning normally.'[5]

Following retrofire, the spacecraft separated as scheduled into three segments: the service module, the descent module (which the crew was occupying) and the orbital module. The braking engines then fired at the correct time on the descent module, which began its re-entry trajectory through the fiery atmosphere. As usual at this stage of the return, all communication with the crew had ceased. After aerodynamic braking took place in the thickening atmosphere, the parachute system was deployed, and just before touchdown the soft-landing engines fired.

The Soyuz capsule landed smoothly in the prearranged zone, and recovery helicopters were quickly on the scene. Rescue crews raced over to the spacecraft and opened the hatch, ready to greet a jubilant crew. Instead, the three cosmonauts were slumped back in their couches as if fast asleep, their faces blue. All three were dead. The rescue team dragged the cosmonauts' bodies from the spacecraft and attempted to revive them, but it was too late.

The frantic scene after rescue crews reach the silent Soyuz-11 spacecraft.

Soyuz Missions 1–11

Flight	Crew	Launched	Landed	Result
Soyuz-1	Vladimir Komarov	23.04.1967	24.04.1967	Parachute failure – pilot killed on impact
Soyuz-2	Unmanned	25.10.1968	28.10.1968	Docking failure with Soyuz-3
Soyuz-3	Georgi Beregovoi	26.10.1968	30.10.1968	Failed to dock with unmanned Soyuz-2
Soyuz-4	Vladimir Shatalov	14.01.1969	17.01.1969	Successful crew transfer from Soyuz-5
Soyuz-5	Boris Volynov Alexei Yeliseyev Yevgeny Khrunov	15.01.1969	18.01.1969	Successfully transferred two crew members to Soyuz-4 by EVA
Soyuz-6	Georgi Shonin Valery Kubasov	11.10.1969	16.10.1969	Intended to observe and film docking of Soyuz-7 and Soyuz-8
Soyuz-7	Anatoly Filipchenko Vladislav Volkov Viktor Gorbatko	12.10.1969	17.10.1969	Failed to dock with Soyuz-8
Soyuz-8	Vladimir Shatalov Alexei Yeliseyev	13.10.1969	18.10.1969	Failed to dock with Soyuz-7
Soyuz-9	Andrian Nikolayev Vitaly Sevastyanov	01.06.1970	19.06.1970	Successful long-duration science mission
Soyuz-10	Vladimir Shatalov Alexei Yeliseyev Nikolai Rukavishnikov	23.04.1971	25.04.1971	First crew to Salyut-1 station but failed to achieve docking
Soyuz-11	Georgi Dobrovolsky Vladislav Volkov Viktor Patsayev	06.06.1971	30.06.1971	Docked with and occupied Salyut-1 but died during return to Earth

Surrounded by garlands of flowers and dressed in dark suits, the bodies of the three cosmonauts lie in state in Moscow's Central House of the Soviet Army prior to their inurnment in the Kremlin Wall Necropolis in Red Square.

Following a lengthy investigation, during which time all crews assigned to future Salyut missions were stood down, the cause was traced to a pressure equalization valve that had opened prematurely beneath Volkov's seat at the time of spacecraft separation, allowing the cabin air to be rapidly sucked out into the vacuum of space. There were indications that the crew had unbuckled and tried desperately to plug the gap, but without protective spacesuits and helmets they were asphyxiated in seconds and died.

Dobrovolsky, Volkov and Patsayev were given a state funeral and their cremated remains were inurned in the Kremlin Wall Necropolis in Moscow's Red Square, near where the ashes of the first cosmonaut, Yuri Gagarin, were similarly inurned.

Despite efforts to keep Salyut-1 in orbit until it could be visited by another Soyuz crew, the deserted station re-entered the atmosphere on 11 October, burning up in a fiery descent over the Pacific Ocean.

8

DÉTENTE IN ORBIT

The loss of the Soyuz-11 crew had shaken the Soviet space programme to the core, but investigators were able to quickly identify and correct the cause. They also acted responsibly, informing NASA officials that it was a hardware fault and in no way attributable to physiological changes suffered by the crew following a lengthy residency aboard the Salyut station. America had been deeply concerned that this might be the case, as NASA was well into preparations to send its astronauts on even longer expeditions to the Skylab space station.

On 27 September 1973, more than two years after the devastating loss of the Soyuz-11 crew, the Soviet Union's piloted spaceflight programme underwent a long-delayed revival with the launch of Soyuz-12 on a spacecraft test flight, carrying Vasily Lazarev and Oleg Makarov into orbit. Their ship was a Soyuz 7K-T, modified to offer a far greater degree of crew safety. In that regard, this and future crews would consist of just two cosmonauts, as there was now a requirement – the first since Voskhod-2 – for all crewmembers to wear protective pressure suits during launch, docking and re-entry. They would spend two days in orbit testing the redesigned spacecraft and its systems before making a safe return to Earth. This test flight was also a precursor to an ambitious binational mission to fly a joint American–Soviet, Apollo–Soyuz, mission in July 1975.

A DECADE EARLIER, on 20 September 1963, President John F. Kennedy caught many of his fellow Americans and Soviet leaders by surprise when, during a speech given at the United Nations, he proposed that the United States and Soviet Union should unite in a combined effort to mount an expedition to the Moon.

'Why, therefore, should man's first flight to the Moon be a matter of national competition?' the young president exhorted, suggesting 'the clouds have lifted a little' in terms of relations between the two nations. 'The Soviet Union and the United States, together with their allies, can achieve further agreements,' he added, 'agreements which spring from our mutual interest in avoiding mutual destruction.'[1]

Both nations had been fully engaged in a politically motivated competition for the high frontier of space since October 1957, when the Soviet Union caught America napping by launching the first Earth-orbiting Sputnik, followed a month later by the massive Sputnik 2, which carried a dog named Laika into orbit. In the two years following the historic first space flight by Yuri Gagarin in 1961, five Russian cosmonauts had been launched into Earth orbit, including the first woman, Valentina Tereshkova. The United States had been forced into playing catch-up, first with two NASA astronauts completing suborbital space shots, followed by four one-man orbital missions of increasing

On 16 November 1963, six days before he was assassinated in Dallas, President Kennedy (right) visited Cape Canaveral, where he was briefed on the Saturn V launch system by Wernher von Braun and NASA Deputy Administrator Robert Seamans (behind von Braun).

President Richard Nixon and Soviet Chairman Alexei Kosygin in Moscow signing the Agreement Concerning Cooperation in the Exploration and Use of Outer Space for Peaceful Purposes, 24 May 1972. Leonid Brezhnev stands behind Nixon's left shoulder.

The U.S. and Soviet ASTP crewmembers. From left: Deke Slayton, Tom Stafford, Vance Brand, Alexei Leonov and Valery Kubasov.

length and complexity. Space had become the undoubted prize in those Cold War years, in which the two nations had faced off with each other during the tense Cuban Missile Crisis of 1962. Yet here was America's president suggesting international cooperation in a mammoth joint expedition to reach the Moon. It happened just two years after he had stood before a joint session of Congress on 25 May 1961, committing his nation – and his nation alone – to 'achieving the goal, before this decade is out, of landing a man on the Moon and returning him safely to Earth'.[2]

Admittedly, previously inflamed temperatures had cooled somewhat since the missile crisis, with more open communication now available between Moscow and Washington. In addition, the signing of a nuclear treaty would take place on 7 October 1963. Kennedy was, therefore, optimistic that the two nations could somehow unite in reaching for the Moon. It could also save both nations billions of dollars to combine their technology and resources. Even Soviet foreign minister Andrei Gromyko saw some merit in Kennedy's proposal, calling it 'a good sign'. But that was as far as it went; Kennedy's overture went unreported in the Soviet press, while public reaction in the United States was less than enthusiastic. The issue does not seem to have been raised again, and the American president was assassinated in Dallas, Texas, just two months after his surprise speech at the United Nations.

According to space historian Francis French,

Shortly after Kennedy's assassination, Congress passed an appropriations bill stating that no money would be given to any international Moon programme. President Lyndon Johnson, newly in office, assertively championed the space race for the rest of the decade, and by the time he left office in 1969, an American Moon landing that year was a virtual certainty.[3]

BY MAY 1972, political attitudes had changed and diplomatic ties were gradually evolving between the United States and the Soviet Union. By then, the race to the Moon had been won by the United States, and preparations for the final crewed mission, Apollo 17, were well under way. There even seemed to be an end in sight to the despised and massively expensive Vietnam War, and President Richard Nixon decided it would be politically expedient to begin mending fences and establish closer ties between the two Cold War nations. Setting aside all the scandals and controversy that would eventually characterize his presidency, Nixon did achieve something of significant and historical note that year when he and the chairman of the Council of Ministers of the USSR, Alexei Kosygin, signed an Agreement on Cooperation in the Exploration and Use of Outer Space for Peaceful Purposes. This

Leonov and Stafford in Houston practising symbolic ceremonies for their upcoming ASTP mission.

NASA cutaway showing the meeting of the astronauts and cosmonauts in the transfer tunnel.

agreement, which also embraced science diplomacy, would lead to the first ever collaborative space mission between the two nations, which became the Apollo-Soyuz Test Project, or ASTP. Early plans called for the rendezvous and docking of an American Apollo and a Soviet Soyuz spacecraft in low Earth orbit in July 1975 and included celebratory visits to each other's spacecraft during two days of joint operations and commemorative activities.

As training had to begin as soon as was practicable, each nation began the process of selecting the cosmonauts and astronauts who would participate in the joint project. The Soviets eventually named veteran cosmonauts Alexei Leonov and Valery Kubasov, with Vladimir Dzhanibekov and Boris Andreyev acting as their back-up crew. The American astronauts assigned as the prime crew were Tom Stafford and Vance Brand, along with Mercury astronaut Donald 'Deke' Slayton, who had recently regained his flight status after a prolonged grounding with a minor medical issue. Their back-ups were named as Alan Bean, Ron Evans and Jack Lousma. Both nations also nominated several cosmonauts and astronauts to work as support crewmembers. The identities of the American crew were announced on 1 February 1973 by Glynn Lunney, NASA's technical director of ASTP. The Russian crew would only be officially named on 24 May 1973 by Konstantin Bushuyev, the ASTP's Soviet project director.

One crucial element of the operation was the development of an androgynous docking module, a form of sophisticated air lock that would allow the Apollo crew to dock with the Soviet Soyuz spacecraft and solve the problem of the two craft operating with their environmental differences. The Apollo command module utilized pure oxygen pressurized at 5 psi, while the Soyuz vehicle maintained a nitrogen/oxygen mix at sea-level pressure. Plans called for the docking module to be carried aloft within the S-IVB stage of the Saturn IB carrier rocket, and once in orbit it would be retrieved by the Apollo crew carrying out a docking manoeuvre in much the same way as lunar modules were extracted from the spent rocket stage during the lunar-landing missions. The two crewed spacecraft could then rendezvous and dock, allowing crew members to become habituated to the other craft's atmosphere during visits.

While the development of testing of flight hardware continued, joint training sessions were also under way. Despite the predicted

Following the historic link-up in orbit, Tom Stafford and Alexei Leonov meet once the hatches have been opened.

Showing no obvious ill-effects from their landing drama, Stafford, Slayton and Brand are shown on the deck of USS *New Orleans*.

protests and deep suspicions expressed (probably with good reason) by both nations, it was deemed crucial for the astronauts and cosmonauts to familiarize themselves with the spacecraft and systems of the 'other side'. In July 1974 the American crews travelled to Moscow and the Yuri Gagarin Cosmonaut Training Centre to acquaint themselves with the Soyuz craft, and, in turn, the Russian crews were based in Houston in September for similar training. Friendships and collaboration quickly evolved between the participants, who would once have been possible combatants in the air, especially as they had to learn what they could of the other country's language. As Alexei Leonov laughingly observed, it was hard enough learning English without having to decipher Tom Stafford's 'Oklahomese'. The two men would become lifelong friends.

ON 15 JULY 1975 the Soviet crew of Leonov and Kubasov were launched into orbit aboard their Soyuz 7K-TM spacecraft (designated Soyuz-19) from the Baikonur Cosmodrome in Kazakhstan. They were followed into orbit 7.5 hours later by the launch of the U.S. crew of Stafford, Brand and Slayton aboard their Apollo CSM-111 from Launch Complex 39 at the Kennedy Space Center in Florida.

Both the Soyuz and Apollo vehicles made orbital adjustments during the following two days, finally bringing the two spacecraft together for a hard docking 229 kilometres (142 mi.) over the Atlantic Ocean as the world watched via live television coverage. Three hours later, having checked that the docking collar was secure and functioning as planned, the hatches between the two spacecraft were opened and the excited crewmembers greeted each other with warm hugs.

Once all the handshakes were over, and before any mission objectives began, it was time for the political aspects of the joint venture, with congratulatory calls from Soviet Communist Party General Secretary Leonid Brezhnev and u.s. president Gerald Ford. There followed an exchange of commemorative gifts and a shared meal before the crews returned to their respective spacecraft and the hatch was closed for the day.

On the second conjoined day, Vance Brand paid a visit to Valery Kubasov in the Soyuz craft, while Alexei Leonov slipped into the Apollo command module, both crews giving TV viewers a 'guided' tour of the two very different cabins. Then it was time to complete some science experiments that were incorporated into an otherwise loose schedule, which would allow the astronauts and cosmonauts to move between the two spacecraft. That afternoon, it was time to make their final speeches and give each other their last handshakes. The hatches were finally secured, and the following morning (19 July), the two spacecraft undocked and withdrew after 19 hours and 55 minutes of joint activities. A second programmed docking was successfully completed and then the two spacecraft separated for the final time.

Soyuz-19 would remain in orbit for a fourth day before re-entering after a highly successful mission, landing under parachutes on the morning of 21 July, less than 12 kilometres (7.5 mi.) from its target zone near the Baikonur Cosmodrome. It had completed a mission of 5 days, 22 hours and 30 minutes. For the very first time, both a Soviet-crewed launch and a Soviet-crewed landing had been televised live.

The Apollo crew would remain in orbit for another three days before they returned to a splashdown in the Pacific Ocean west of Hawaii on 24 July, after a mission lasting 9 days, 1 hour and 28 minutes.[4] There was unexpected drama for the Apollo crew during re-entry when they were accidentally exposed to toxic nitrogen tetroxide fumes, caused by reaction control system (RCS) oxidizer venting from the spacecraft and re-entering the cabin through an air intake that was mistakenly left open. Vance Brand fell unconscious for a short time while Stafford hurriedly retrieved their emergency oxygen masks and put them on. Following their safe retrieval, the three astronauts were transported to a hospital in Honolulu, where they all made a complete recovery.

The joint mission in July 1975 would be regaled as an outstanding demonstration of détente, marking the end of the Cold War space race. 'The legacy of Apollo-Soyuz is the foundation [it laid],' General Tom Stafford recalled for the *Moscow Times* newspaper in July 2015 at a celebration to mark the fortieth anniversary of the historic mission, which became universally known as the 'handshake in space'. He said it demonstrated 'that two countries, the two superpowers in the world, with different languages, different systems of measurement, and two different political systems could work together for a common goal and do it successfully, [thereby laying] the foundation for future work in space'.[5]

The ASTP mission represented the first time the two Cold War space programmes had united in a common cause, ushering in a whole new era of outer space cooperation that lives on today in the form of the ISS.

9

SPACE STATION MIR

Ostensibly identified as Salyut-2 in order to disguise its real purpose as an Almaz military reconnaissance facility, the next Soviet space station was launched on 3 April 1973. However, it suffered serious systems and pressurization problems before the first occupying crew could be sent to a link-up in orbit. Badly damaged and rendered untenable, Salyut-2 would eventually tumble out of orbit and burn up over the Pacific Ocean during its premature re-entry on 28 May 1973.

Thrust into orbit on 25 June 1974, a second top-secret Almaz military station, Salyut-3, was followed into space eight days later by the crew of veteran cosmonaut Pavel Popovich and flight engineer Yuri Artyukhin. After a successful hard dock, they occupied the station for two weeks, carrying out what TASS reported to be mostly medical tests and experiments, but their schedule was obviously augmented by highly classified work and observations.

A second crew was launched to the space station aboard Soyuz-15 on 26 August 1974, but their Igla automatic docking system failed and they were unable to complete a rendezvous and docking with Salyut-3. As a result, their flight was curtailed and they were brought home early. Salyut-3 would be deorbited five months later, in January 1975, without ever playing host to another crew.

Soyuz-16 was the next mission to launch from Baikonur in December 1974, but this was described as an orbital test of the autonomous docking system, to be used on the upcoming Apollo-Soyuz Test Project flight. The morning after Christmas Day 1974, the Soviets launched their fourth space station into orbit, identified as Salyut-4. This time it was classed as a civilian rather than military facility. Two further Soyuz crews (17 and 18) would occupy and work aboard Salyut-4 over periods of 30 and 63 days, respectively, before it began to fall into disrepair and was abandoned. It became obsolete with the launch of Salyut-5 on 22 June 1976. Another crew had earlier been scheduled to dock with Salyut-4, but they suffered a launch abort and their flight was abandoned. The failed flight was subsequently redesignated Soyuz-18A. After more than two years in space, and during which it needed its altitude to be boosted several times, Salyut-4's orbit was rapidly decaying. Its job done, ground controllers activated the station's propulsion system, resulting in a fiery re-entry over the Pacific Ocean on 2 February 1977, during which it was destroyed by intense frictional heat.[1]

Salyut-5, which began orbital service in June 1976, was the third and final military Almaz space station, created as a series of top-secret observation and reconnaissance facilities. Among other features, a large telescope almost 1 metre (3.3 ft) in diameter was mounted on the floor of the station, which allowed crew members to photograph such ground installations as military airfields and missile launch complexes. The station would be visited by two crews, Soyuz-21 and Soyuz-24. Soyuz-22's mission did not include a link-up with Salyut-5, while Soyuz-23 failed to dock with the space station owing to yet another fault in the troubled Igla docking system. It was reported that the station had begun running critically low on propellant, so a fourth boarding mission that had been planned was abandoned. Salyut-5 was deorbited on 8 August 1977.[2]

WHILE THE FIRST Soviet space stations met with limited success and hosted a small number of visiting crews, Salyut-6 and -7 proved to be exceptional research and testing facilities. The real advantage lay in the fact that they were both fitted with two docking ports, one at either end, allowing two Soyuz crews to be docked to the station at the same time. Unfortunately, the first crew to attempt a docking with Salyut-6 (Soyuz-25) failed to complete it and had to return home.

Among the many crews inhabiting Saylut-6 during its orbital life, from 1977 to 1981, were eight international pilots from Soviet bloc countries who flew to the station as 'guest cosmonauts' along with an experienced Soviet commander in the highly successful Interkosmos programme. The first such flight carried Soviet cosmonaut Alexei Gubarev along with Vladimir Remek from Czechoslovakia, with each

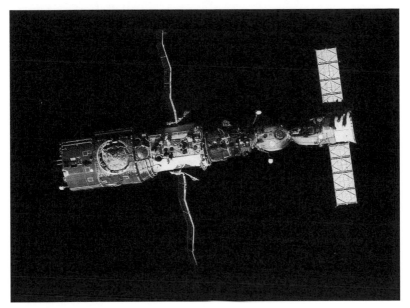

Salyut-6 with a Progress supply craft docked, photographed by the Soyuz T-4 crew.

of the eight internationals spending a week aboard the orbiting laboratory. Another mission flying a ninth crewmember from Bulgaria (Soyuz-33) had to be aborted and returned to Earth following a docking failure with the station.

Between the first crew arrival at the station (Soyuz-26) on 11 December 1977 and the final crew departure (Soyuz-40) on 22 May 1981, Salyut-6 played host to eighteen occupying crews and twelve unmanned Progress supply spacecraft. It was de-orbited on 29 July 1982 and replaced by the vastly updated Salyut-7, already positioned in low Earth orbit.

SALYUT-7 – 14.4 METRES (47.2 ft) long, 4.15 metres (13.62 ft) in diameter and a total mass of 19,920 kilograms (44,000 lb) – was launched from the Baikonur Cosmodrome on 19 April 1982 and would replace the obsolete Salyut-6. It was destined to become the last space station in the Salyut programme, as it was superseded and replaced in orbit by the modular Mir (Peace) station. First occupied on 14 May 1982 by the Soyuz T-5 crew of Anatoli Berezovoy and Valentin Lebedev, Salyut-7 would receive twelve crewed and fifteen unmanned spacecraft until the final crew (Soyuz T-15) departed on 25 June 1986, leaving it adrift in orbit.

The crew of the first Interkosmos international mission: Soviet commander
Alexei Gubarev (left) and Czech pilot Vladimir Remek.

Continuing the theme of internationally sponsored visitors,
Salyut-7 would play host to guest crewmembers from France (Jean-
Loup Chrétien) and India (Rakesh Sharma). Another notable arrival
aboard Soyuz T-7 was Svetlana Savitskaya, the first Soviet woman
launched into space since Valentina Tereshkova nineteen years earlier.
Svetlana would subsequently make a second flight to Salyut-7 in July
1984 aboard Soyuz T-12. The Soviets had been alerted to the fact that
NASA astronaut Kathy Sullivan was training to become the first woman
to perform an EVA during the STS-41G Space Shuttle mission. As a result,
Svetlana's occupancy of the station included beating Sullivan to this
space 'first' by completing her own EVA and conducting some elemen-
tary welding work outside of Salyut-7, along with her T-12 crewmembers,
Vladimir Dzhanibekov and Igor Volk. Their mission ended with a safe
touchdown on 29 July 1984.

One of the more remarkable stories concerning Salyut-7 occurred in February 1985, when the then-vacant space station 'died' after it lost power and shut down. Despite several attempts by ground controllers to restore the power supply, their efforts were in vain. The station had also begun slowly tumbling in orbit. The launch of the replacement Mir station was still at least a year away, so rather than simply abandoning Salyut-7, it was decided to launch a salvage mission, even though it presented many risks to the cosmonauts. Given that the station was unstable, a failed docking could have damaged the Soyuz spacecraft, stranding the crew in space. It bore some comparisons with the rescue mission of Apollo 13, even to the spacecraft bearing the designation T-13. The crew of Vladimir Dzhanibekov and Viktor Savinykh lifted off on 6 June 1985, and fortunately the docking – albeit difficult and dangerous – was successfully completed.

Checks showed that the station's life-support system was operating at normal interior levels, so there was no need to don their spacesuits before proceeding inside. However, knowing from ground data that the temperature inside Salyut-7 was sub-zero, the two cosmonauts put on thick woollen caps and heavy coats before entering the station. The freezing interior was pitch black; ice had formed on the walls and instruments, and the water supply was frozen. Under these terrible and hazardous conditions, the two cosmonauts spent

The Soyuz T-7 crew on board Salyut-7: Leonid Popov (left), Aleksandr Serebrov and Svetlana Savitskaya (second Soviet woman in space). On a later mission, Savitskaya became the first woman to conduct an EVA.

the next ten days slowly coaxing the crippled station back to life. They managed to connect some batteries to the exterior solar panels and align the station with the Sun. Ever so slowly, power began to seep back into Salyut-7, and when it was sufficient, Dzhanibekov and Savinykh were able to bring the station's systems back online. It was an heroic effort, allowing that crew and others to safely occupy the ageing station, while uncrewed Progress spacecraft could now dock at the station's vacant end port, bringing fresh supplies for the resident crews.[3]

There may have been plans to keep Salyut-7 in orbit, perhaps for later occupancy, as it was boosted into an unusually high altitude of around 480 kilometres (300 mi.), which would have kept it circling the planet for several more years. Despite this, the abandoned station fell victim to high, unforeseen solar activity that created increased atmospheric drag, speeding Salyut-7's orbital decay and end. After nine

Viktor Savinykh and Vladimir Dzhanibekov, the T-13 crew tasked with salvaging the troubled Salyut-7, shown aboard the ice-cold space station wearing warm headgear.

years in orbit, and five years since its last occupancy, Salyut-7 finally re-entered Earth's atmosphere, burning up and blazing a fiery trail across the skies over Argentina.[4]

FOR ONE OF THE TWO major spacefaring nations, 1986 would begin with a catastrophic tragedy, and for the other it would mark the beginning of a new era of space exploration. On 28 January Space Shuttle *Challenger* was lost 73 seconds after lift-off, together with its seven crewmembers. On the other side of the world, the Soviet Union was preparing to launch the core element of a new, innovative space station.

On 19 February a Proton rocket tore a path into the skies over Kazakhstan, carrying the core element of the next-generation Soviet space station into orbit. Western observers fully expected the next laboratory to be named Salyut-8, but once it had settled into orbit and details were announced, the new station was referred to as 'Mir'. The word connotes 'village' or 'community', but it can also be translated as 'peace', which was the name formally given to the station during public bulletins.

Mir would become a larger and far more complex craft than any of its predecessors, continuing the Soviet Union's mission to study the physiological and even mental effects of prolonged space flight on the human body. It would allow for a wide range of research tests and experiments, Earth studies and astronomical observations as well as routine maintenance chores. It had been designed in such a way that once established and crewed, a phased operation would begin in which a number of modules would be incrementally launched and attached to one of Mir's docking ports. Over time, this would greatly increase the size and capabilities of the orbiting laboratory.

On 13 March 1986 the cosmonauts Leonid Kizim and Vladimir Solovyov became the first crew launched to the Mir space station. After spending 55 days checking and activating Mir's system and unloading two unmanned Progress supply craft, they undertook a planned mission to undock and fly their Soyuz T-15 spacecraft to a link-up with the abandoned Salyut-7 craft. Following a 29-hour flight, they docked with Salyut-7. They would spend eight weeks collecting equipment and experiments to take back to Mir and performed two spacewalks. On 25 June they undocked from Salyut-7 and completed a safe docking back at Mir, where they transferred the retrieved items. It would be the only time a crew had flown from one space station to another.

The Mir space station.

Over a four-year period beginning on 31 March 1977, three purpose-built modules would be launched aboard Proton rockets to an eventual rendezvous with Mir. The first of these modules was designated Kvant-1 (Quantum-1), which contained instruments for material-science experiments and astrophysical observations. After some worrying trajectory mishaps, a soft docking with the station was eventually achieved on 9 April, but something was preventing a hard dock. The then-resident crew of Yuri Romanenko and Aleksandr Laveikin had arrived at the station in February aboard the first of a new series of Soyuz spacecraft, designated the TM variant. They were required to carry out a contingency EVA to determine why the hard dock had not taken place. Making their way to the docking collar, they found that a rubbish bag had somehow become stuck in the collar during the departure of the unmanned Progress 28 supply craft. They removed the obstruction, and once the cosmonauts were safely back inside Mir, the docking procedure was carried out successfully.

ON 15 NOVEMBER 1988 a delta-winged spacecraft named Buran (Snowstorm) touched down on a concrete runway 12 kilometres (7.5 mi.) from the Baikonur Cosmodrome in Kazakhstan. In appearance, it was almost identical to NASA's Space Shuttle orbiters – even down to having 38,000 individually crafted thermal protective tiles glued to its exterior – but unlike its American counterpart, there was no human presence on board. The fully automatic landing, following two pre-planned orbits lasting 205 minutes, was flawless. Buran, like the space shuttles it so closely resembled in shape and dimensions, was meant to usher in an exciting new dimension and purpose to the Soviet space programme. However, it proved so calamitously expensive that the maiden test flight – a computer-guided precursor to a later crewed mission – would become its first and last venture into space.

Launched astride a non-recoverable Energia rocket, and unlike the U.S. space shuttles, Buran did not have main engines mounted within its tail as an integral part of the launch system. This meant it could carry heavier payloads into orbit than the NASA orbiters, although it did have two jet engines mounted on its rear end. These would give the later pilots the opportunity to manoeuvre Buran during the landing phase, allowing them to make a second attempt in the event of adverse conditions, such as unexpectedly strong crosswinds, whereas NASA's shuttle pilots, flying what was basically an overweight glider,

The Buran shuttle photographed on the Baikonur launch pad mounted on an Energia rocket.

did not have that facility. They were committed to a one-shot-only unpowered landing.

Buran was lauded as a great triumph of Soviet engineering and technology, and plans were drawn up for a second unpiloted orbital test flight lasting up to twenty days. Even as specifically selected Buran pilots were undergoing familiarization training with simulators, two critical factors saw the Buran programme irreversibly terminated. The first was the dissolution of the Soviet Union near the end of 1991, and the second was the spiralling and unsustainable cost of maintaining the programme. There was simply no money for it to continue, as a heavily cash-strapped nation struggled with the demise of its once powerful dominion.

Now regarded as nothing more than an unwanted, overly expensive test vehicle, Buran became the property of the newly formed republic of Kazakhstan. Undeservedly consigned to a leaky hangar at the Baikonur launch site, it would languish there, mounted horizontally on top of a mock-up of an Energia rocket. Then, on 12 May 2002, the heavily waterlogged hangar roof collapsed during a fierce rainstorm, crashing down and crushing Buran while also killing eight workers who were attempting to repair the roof. It was a sad and poignant end to the once magnificent winged spacecraft and the innovative space programme it had once heralded.

IN DECEMBER 1989 the Kvant-2 module was added to Mir, featuring an EVA airlock, an instrument/cargo compartment and an instrument/experiment compartment. The final module to be attached, named Kristall (Crystal), was launched on 31 May 1990, docking with Mir on 10 June. Its principal purpose was to develop biological and materials production technologies in the microgravity of space. Several years later, two new modules would be added. There was Spektr (Spectrum) launched in May 1995, with four solar arrays that would greatly increase the station's electrical power, as well as equipment for Earth observation. Then, Priroda (Nature), the seventh and final module, was added in late April 1996. Priroda would allow research into Earth resource experiments through developing and verifying remote sensing methods.

Throughout its fifteen-year life span in orbit, the Mir station would acquire a somewhat mixed legacy. It witnessed the start of a truly collaborative, international effort. This would enable NASA's space shuttles to rendezvous and dock with the Soviet station and host not only Russian and Ukrainian cosmonauts, but spacefarers from the

The first and only landing of the Buran shuttle on 15 November 1988, accompanied by Buran cosmonaut Igor Volk flying an MiG-25 chase plane.

United States, France, the United Kingdom, Japan, Syria, Bulgaria, Afghanistan, Austria, Germany, Slovakia and Canada. In addition, over the years, the now-massive station served as a microgravity research laboratory in which rotating crews performed numerous experiments in biology, human biology, physics, astronomy, meteorology and spacecraft systems. Many individual records were created (and subsequently beaten) for the most time spent in space. The champion was Soviet cosmonaut Valery Polyakov, who, from January 1994 to March 1995, set a record of just under 438 continuous days living and working aboard Mir.

Then there was cosmonaut Sergei Krikalev. He began his second flight to Mir on 18 May 1991 as a Soviet citizen and was still working as a flight engineer aboard the space station during the collapse and dissolution of the Soviet Union. Krikalev was informed he would have to extend his stay while the politics, finances and future of the space programme were being thrashed out on the ground. He finally returned having accumulated 311 days in orbit, his Soyuz TM-13 spacecraft touching down near the city of Arkalyk, Kazakhstan – now an independent state. He was weak, pale and perspiring, but happy to be back on solid ground once again. However, the nation to which he had belonged at launch no longer existed, and in the Western media he was given the apt tongue-in-cheek sobriquet of 'the last Soviet citizen'.

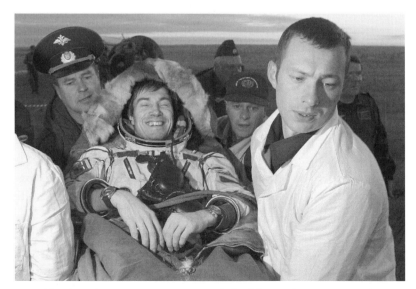

Record-breaking 'last Soviet citizen' Sergei Krikalev following his return to Earth on 25 March 1992 after spending 311 days in space.

The seven NASA astronauts who would take up successive occupancies aboard Mir. Back row (from left): Andy Thomas, Shannon Lucid and Mike Foale. Front row: Norm Thagard, John Blaha, Jerry Linenger and David Wolf.

On the debit side of Mir's legacy, the station also endured a number of critical problems, including power failures, a near-disastrous fire that almost caused an emergency evacuation by the crewmembers and a life-threatening collision with an unmanned supply craft.

AN UNCONTAINED FIRE aboard an orbiting spacecraft is a terrifying prospect, and this life-threatening scenario erupted aboard Mir on 23 February 1997. NASA astronaut and medical doctor Jerry Linenger was the fourth American to live and work aboard Mir, and he recalled working at a computer terminal in another module when the fire broke out. A solid-fuel oxygen generator canister had burst into flames. 'It was actually a backup oxygen generator that caused it,' Linenger stated. 'It sparked what was supposed to be an exothermic reaction and instead, it's on fire. Three-foot flames, blowtorch-like in intensity. Sparks flying off the end of it; melting metal.'[5]

Alarms were sounding, and smoke began belching throughout the station. Linenger grabbed an emergency oxygen mask and put it on, but it wasn't working, so he found another, which was fortunately operational. Fire extinguishers were rushed to the source of the fire and operated, although someone had to stand behind the operator to stop them from floating backwards owing to a lack of gravity and the reactive principle involved in Newton's third law of motion. The next problem was that the fire and smoke were blocking the six crewmembers' access to one of the two Soyuz spacecraft docked at either end of Mir, which could be used in an emergency evacuation from the space station. As each Soyuz only carried three people, it was a dire situation. 'Without getting that fire out, there was no way to get to one of the Soyuz capsules,' Linenger commented, 'and we were planning on evacuation if needed . . . We went through four fire extinguishers and finally got it out after about fourteen minutes. There was a total blackout and you couldn't see the five fingers in front of your face.'

Once the smoke had slowly dissipated, Linenger set about examining his five fellow crewmembers, but fortunately none of them had suffered too badly from smoke inhalation. It had been a very close call.[6] Four months after the fire, with Jerry Linenger and other crewmembers now departed and replaced aboard Mir, the station would once again be involved in a life-or-death situation. This time the danger came when a robotic Progress cargo ship slammed into Mir during a docking test.

On 25 June the station's Russian commander, Vasily Tsibliev, was conducting a scheduled manual docking systems test of the 7-tonne Progress M-34 craft, which had brought supplies, including food, water and oxygen, to Mir two months earlier, as well as three replacement fire extinguishers. It had also transferred some spare propellant into the station's tanks. Since then, the ship had remained docked at Mir's rear port, loaded with rubbish for its eventual undocking and destructive re-entry. These early Progress craft were routinely regarded as expendable; they were not considered important enough to warrant the massive expense of fitting them for a protected re-entry and organizing their subsequent recovery.

It would prove a difficult task for Tsibliev. He had no solid data about the range or speed of the approaching cargo ship to work with, apart from what he could deduce using a stopwatch and hand-held rangefinder. In the end, that would all prove totally useless, as Progress M-34 began to glide out of sight, its path obscured by Mir's solar panels.

As it approached Mir, the bus-sized craft suddenly accelerated out of control, bypassed the intended docking port and crashed into the complex array of modules that had been added to Mir's core component. It then bounced off the Spektr laboratory module into the solar arrays and a radiator. All three crewmembers were shocked by the sudden and severe impact but were otherwise uninjured. However, the collision had punched a hole in Spektr's laboratory module, where British-born NASA astronaut Michael Foale was working.

Fortunately, the hole was less than an inch in diameter. As Mir's precious oxygen whistled out into the vacuum of space, Foale, Tsibliev and Aleksandr Lazutkin fought to plug the leak. In order to do this they had to slice cables carrying power from Spektr's solar panels, cutting power to the station by about half. Nevertheless, they would count themselves lucky; had the hole been any larger, an uncontrollable decompression would have rapidly suffocated the three men, who were not wearing spacesuits. As journalist Kathy Sawyer later reported in the *Washington Post*:

> In the aftermath of the crash, the crew worked inside a darkened, partially shut-down facility, using as little electricity as possible while maneuvering to recover lost power. Officials said their most pressing concern was that Mir has lost 50 percent of its total potential power generating capability in the accident, which could jeopardize their ability to run the life-support system and positioning control.[7]

The crew repaired what they could and manoeuvred the slowly tumbling Progress M-34 well away from the station before it was eventually deorbited on 2 July. When the next Soviet crew replaced Tsibliev and Lazutkin, they carried with them items and instructions necessary to repair most of the remaining damage. It was becoming increasingly obvious that the ageing space station was becoming a liability, although crews would continue to be launched to Mir until 4 April 2000, with the Soyuz TM-30 crew of Sergei Zalyotin and Aleksandr Kaleri. On 16 June they closed the hatch on Mir for the final time and left it a drifting, derelict ghost ship.

In March 2001 the decision was made to deorbit Mir. It had suffered through (and mostly recovered from) a number of serious problems that had disrupted its service as an orbiting research laboratory, but its systems and hardware had now become severely outdated and costly to maintain. There was no longer any money available to keep Mir operational, particularly as Russia had recently committed to participate in the International Space Station programme, leaving insufficient funds to support both projects. Mir would therefore have to be decommissioned.

The re-entry process began when ground controllers fired Mir's thrusters, which gradually lowered the 143-tonne station's orbit, allowing gravitational forces to take increasing control. On 24 January 2001, with the Progress MI-5 supply craft still attached, Mir finally succumbed to the inescapable clutches of gravity and began to re-enter the atmosphere. After fifteen years in Earth orbit and a journey of 3.38 billion

Crews to Salyut-3/Almaz Space Station

Flight	Crew	Launched	Landed	Result
Soyuz-14	Pavel Popovich Yuri Artyukhin	03.07.1974	19.07.1974	Occupied Salyut-3 for 16 days
Soyuz-15	Gennady Sarafanov Lev Demin	26.08.1974	26.08.1974	Failed to dock with Salyut-3

(Soyuz-16 was an ASTP test mission, not planned to dock with Salyut-3. Crew: Anatoli Filipchenko and Nikolai Rukavishnikov; 02.12.1974–08.12.1974)

Crews to Salyut-4 Space Station

Flight	Crew	Launched	Landed	Result
Soyuz-17	Alexei Gubarev Georgi Grechko	11.01.1975	09.02.1975	Occupied Salyut-4 for 30 days
Soyuz-18A	Vasily Lazarev Oleg Makarov	05.04.1975	05.04.1975	Launch abort, flight number redesignated
Soyuz-18	Pyotr Klimuk Vitaly Sevastyanov	24.05.1974	26.07.1975	Occupied Salyut-4 for 63 days
Soyuz-20	No crew	17.11.1975	16.02.1976	Supply and test craft

Salyut-6: Visiting Soyuz Crews

Flight Up	Launch Date	Crew	Days Duration	Flight Down	Landing Date
Soyuz-26	10.12.1977	Yuri Romanenko Georgi Grechko	96.42	Soyuz-27	16.03.1978
Soyuz-27	10.01.1978	Vladimir Dzhanibekov Oleg Makarov	5.96	Soyuz-26	16.01.1978
Soyuz-28	02.03.1978	Alexei Gubarev Vladimir Remek (Czechoslavakia)*	7.93	Soyuz-28	10.03.1978
Soyuz-29	15.06.1978	Vladimir Kovalyonok Aleksandr Ivanchenkov	139.62	Soyuz-31	02.11.1978
Soyuz-30	27.06.1978	Pyotr Klimuk Miroslaw Hermaszewski (Poland)*	7.92	Soyuz-30	05.07.1978
Soyuz-31	26.08.1978	Valery Bykovsky Sigmund Jähn (German Dem. Rep.)*	7.87	Soyuz-29	03.09.1978
Soyuz-32	25.02.1979	Vladimir Lyakhov Valery Ryumin	175.02	Soyuz-34	19.08.1979
Soyuz-33	10.04.1979	Nikolai Rukavishnikov Georgi Ivanov (Bulgaria)	47.01	Soyuz-33	12.04.1979
Soyuz-36	26.05.1980	Valery Kubasov Bertalan Farkas (Hungary)*	7.87	Soyuz-35	03.06.1980
Soyuz T-2	05.06.1980	Yuri Malyshev Vladimir Aksyonov	3.93	Soyuz T-2	09.06.1980
Soyuz-37	23.07.1980	Viktor Gorbatko Pham Tuan (Vietnam)*	7.86	Soyuz-36	31.07.1980
Soyuz-38	18.09.1980	Yuri Romanenko Arnaldo Tamayo Mendez (Cuba)*	7.86	Soyuz-38	26.09.1980
Soyuz T-3	27.11.1980	Leonid Kizim Oleg Makarov Gennady Strekalov	12.8	Soyuz T-3	10.12.1980
Soyuz T-4	12.03.1981	Vladimir Kovalyonok Viktor Savinykh	74.73	Soyuz T-4	26.05.1981
Soyuz-39	22.03.1981	Vladimir Dzhanibekov Jugderdemidiin Gurragchaa (Mongolia)*	7.86	Soyuz-39	30.03.1981
Soyuz-40	14.05.1981	Leonid Popov Dumitru Prunariu (Romania)*	7.86	Soyuz-40	22.05.1981

* Interkosmos crew member

kilometres (2.1 billion mi.), during which it hosted 104 space explorers from many nations, the doomed station became a massive, disintegrating fireball. The remaining pieces crashed into a pre-determined area in the South Pacific Ocean, southeast of Fiji. It was the fiery end of an amazing era in the history of human space exploration.

10

THE SOYUZ LEGACY

In the predawn darkness of 21 July 2011, Space Shuttle *Atlantis* swooped down over Florida and mission commander Chris Ferguson lined up for a landing on Runway 15 of the Kennedy Shuttle Landing Facility. At 5:57 a.m. EDT, *Atlantis* touched down and rolled to a halt, with Ferguson proudly stating, 'Mission complete, Houston. After serving the world for over thirty years, the space shuttle has earned its place in history. It's come to a final stop.'[1]

Sadly, the Space Shuttle's complexity had resulted in the programme's greatest failures: the loss of *Challenger* in 1986 and the disintegration of *Columbia* while landing in 2003. They were two tragic accidents that took the lives of fourteen crewmembers in all. The space shuttle had otherwise served its purpose over three decades of operations, but the time had come to move on to a new generation of spacecraft that would go beyond missions to low Earth orbit, and following the loss of *Columbia*, NASA had prepared a timetable for the retirement of the Shuttle fleet. Mission STS-135 was the final chapter in the Space Shuttle story, drawing the curtain on an unforgettable era in spaceflight history.

THE RETIREMENT OF NASA's Space Shuttle fleet left Russia's venerable, far-less-sophisticated Soyuz spacecraft as the only available means of transporting crewmembers of all nations to the ISS. It was time to take advantage of the situation, with Russian space agency Roscosmos smugly declaring, 'From today, the era of the Soyuz has started in manned space flight, the era of reliability.' However, the agency expressed its admiration for the Shuttle programme, which

The Soyuz TMA-2 spacecraft is rolled out to the Baikonur launch pad.

Showing the cramped conditions in a Soyuz spacecraft are the Soyuz TMA-13M international crew of (from left) Alexander Gerst (Germany, European Space Agency), Soviet commander Maksim Surayev and Gregory Wiseman (NASA).

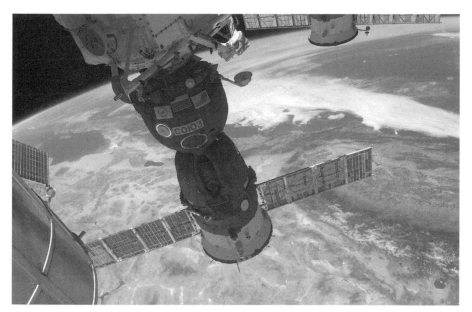

The Soyuz TMA-13M spacecraft is shown docked to the ISS in May 2014, photographed flying over California's Mojave Desert.

it said had delivered payloads to space that were indispensable in the construction of the ISS. 'Mankind acknowledges the role of American space ships in exploring the cosmos.'[2]

By this time, Russia had begun launching a lighter, modified TMA-M version of Soyuz, with digital rather than analogue computers, although each flight still relied on a parachute landing at the end of each mission. 'Reliability and not to mention cost efficiency,' Roscosmos pointed out.[3]

In the nine-year period between the retirement of the Space Shuttle in 2011 to the first crewed mission aboard SpaceX's Crew Dragon in 2020, Soyuz TMA-M and later the MS spacecraft provided the sole means of ferrying crews, which included members of several nations (Japan, France, Italy, Germany, Canada and the UAE), to and from the ISS.

What is truly remarkable about the spherical Soyuz spacecraft (and their similarly named carrier rocket) is how little they have actually changed between the 1960s and today, more than fifty years since first flying into space. There were early technical problems that resulted in the tragic fatalities of Vladimir Komarov aboard Soyuz-1 in 1967 and the three-man Soyuz-11 crew in 1971, but since then, the Soyuz spacecraft

Cosmonauts Oleg Kotov and Sergey Ryazanskiy pose with the Olympic Torch during an EVA outside the ISS on 9 November 2013. The torch was later used to light the Olympic Cauldron in Sochi, Russia.

and rocket have undoubtedly proven themselves to be the most frequently used, reliable and economical means of reaching orbit and linking with the ISS. Roscosmos once estimated that for the cost of a single Shuttle mission, they could launch twenty Soyuz spacecraft.

The roominess of the Space Shuttle was definitely a distinct advantage over the cramped Soyuz capsule, but its very complexity would often cause problems for the crews. Italy's Major Paolo Nespoli from the European Space Agency (ESA) has flown on both spacecraft; aboard Shuttle *Discovery* on mission STS-120 and twice on Soyuz expeditions to the ISS. 'We could learn a lot from the Russians that sometimes when you do less, less is better,' he declared. 'The Shuttle is super complicated, and then things start breaking, they don't work, complexity costs.'[4]

The main disadvantage in the Soyuz spacecraft is a distinct lack of room. 'It's much too small and tight,' said Dutch astronaut André Kuipers, also from the ESA. 'Especially the left and right seats where the tall Europeans and Americans sit. The very first time I went into the

Soyuz training capsule, I saw my American colleague taking painkillers. I asked him why he did that, and he said "You'll find out!" and indeed I had a lot of problems with my knees – it's very uncomfortable.'[5]

Landing by parachute can also provide a dramatic ending to a mission for the crew. 'It's a huge car crash at best, very violent,' recalled Nespoli. 'You look at some of the hardware in the capsule and think, "Wow, we're back in the '50s!"' Nevertheless, the philosophy is: if it ain't broke, don't fix it. 'The vehicle works, it's effective, does exactly what it's supposed to do,' Nespoli said of his Soyuz experiences.[6]

On 14 October 2020 the Soyuz MS-17 spacecraft lifted off from the Baikonur Cosmodrome, achieved orbit nine minutes later and, in an 'ultra-fast' approach to the ISS, achieved a docking on the second orbit, just three hours after leaving the launch pad. In addition to Roscosmos cosmonauts Sergei Ryzhikov and Sergei Kud-Sverchkov, NASA astronaut Kate Rubins was the third crewmember on board. She had been added to the crew in May, when NASA announced it was buying a final seat from Roscosmos for U.S.$90.25 million. She was born in 1978, eleven years after the first Soyuz flew into space.

Unless things change, with commercial crew missions now being launched to the ISS, Kate Rubins may represent the last time NASA will have to pay to place one of its own astronauts on a foreign spacecraft – at least for the foreseeable future.[7]

IN NOVEMBER 2020 a dramatic announcement was made involving a joint project of the Russian State Corporation Roscosmos, Russian TV Channel One and the Yellow, Black and White film and television studio. A competition was being launched to select a suitable professional Russian actress willing to fly to the International Space Station on the 147th crewed flight of a Soyuz spacecraft to participate in the making of a feature film. In announcing the competition the General Director of Roscosmos, Dmitry Rogozin, stated that the Soyuz MS-19/Expedition 66 mission to the ISS would support the making of a film, titled *Vyzov* (Challenge), aboard the orbiting outpost. The actress would be accompanied on the Soyuz flight by the film's director, Klim Shipenko, and experienced mission commander Anton Shkaplerov, who would double as flight engineer.

It would mark the first time in 21 years that a Soyuz craft was launched with an entirely Russian crew on board, and the first time that two of the three seats in a Soyuz craft would be occupied by

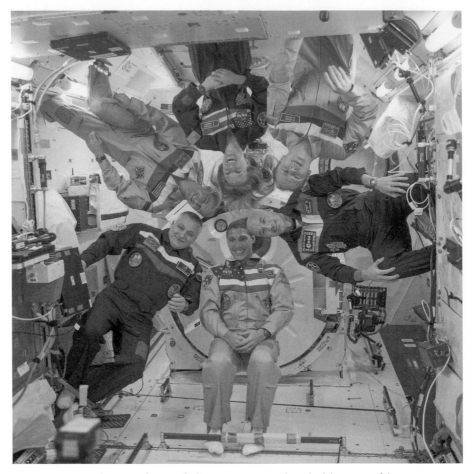

Six crewmembers pose for an inflight crew portrait in the Kibo laboratory of the
ISS. Clockwise from lower left: Commander Fyodor Yurchikhin (Russia), Sergey
Ryazanskiy (Russia), Karen Nyberg (NASA), Oleg Kotov (Russia), ESA astronaut
Luca Parmitano and Michael Hopkins (NASA).

tourists. Sales of these privately funded flights resumed after Roscosmos
lost its monopoly on flights to the space station after SpaceX began
ferrying astronauts to the station in its Dragon spacecraft.

The announcement of the Russian film came soon after news broke
that American film star Tom Cruise was also planning to film a
sequence for an upcoming action film aboard the ISS, along with producer Doug Liman and an unnamed actress. This flight would only
take place after the Russian film crew had returned safely. After a
selection process had been completed, four female Russian actors were
still in the running for the flight, with two more to be eliminated in

May 2021. The remaining two would begin mission training at the Yuri Gagarin Cosmonaut Training Centre, at all times under evaluation while vying for a position on the primary crew, with the runner-up to serve as a backup crew member along with associate director Alexei Dudin and cosmonaut commander Oleg Artemyev.[8] As Dmitry Rogozin explained: 'This is a kind of space experiment. The actress selected by a competition and a medical commission will need to perform the functions of a cosmonaut-researcher and become a full-fledged member of the crew.'

Director Klim Shipenko revealed that the plot of the film involved a male cosmonaut suffering a cardiac arrest while performing a space-walk outside the ISS. He survives, but requires urgent surgery, so he is able to pilot a Soyuz spacecraft back to Earth. Zhenya, a female cardiac surgeon, has only weeks to train for a Soyuz flight to the space station. The resident Russian crew of cosmonauts Oleg Novitsky and Pyotr Dubrov, in orbit since April 2021, would be involved to some extent in the film, and possibly other occupying crew members from NASA (Mark Vande Hei, Shane Kimbrough and Megan McArthur), the European Space Agency (ESA, Thomas Pesquet) and the Japan Aerospace Exploration Agency (JAXA, Akihiko Hoshide).

Eventually the four actress candidates were named as Alyona Mordovina, Yulia Peresild, Sofya Arzhakovskaya and Galina Kairova, and following further tests and interviews Mordovina and Peresild were named on 19 May as the two finalists.[9] The MS-19 flight was tentatively scheduled for lift-off on 5 October, with Peresild and Shipenko returning aboard the Soyuz MS-18 spacecraft twelve days later.

On 10 September 2021 the orbit of the ISS was raised in preparation for the arrival of the Soyuz MS-19 spacecraft and its crew. Anton Shkaplerov would remain on board the ISS until sometime in 2022.

The Soyuz MS-19 mission lifted off on schedule on 5 October before docking with the ISS after a two-orbit rendezvous. The docking proved troublesome, with Shkaplerov having to take over manual control to complete the link-up. Following checks they opened the hatches and moved into the station, meeting the resident crewmembers and preparing to spend the next twelve days filming. On 17 October 2021 the actress and director safely returned aboard the Soyuz MS-18 spacecraft, along with Russian commander Oleg Novitsky.

Earlier, in May 2021, it was announced that two Japanese space tourists were to begin pre-flight preparations for the Soyuz MS-20 flight

The crew of Soyuz MS-19 (from left): Russian actress Yulia Peresild, mission commander Anton Shkaplerov and film director Klim Shipenko.

to the ISS in December that year. Billionaire Yusaku Maezawa and his assistant, film producer Yozo Hirano, would launch along with space-craft commander Roscosmos cosmonaut Aleksandr Misurkin, with Aleksandr Skvortsov training as his backup for the twelve-day mission. The two men began their pre-flight training in June at the Yuri Gagarin Cosmonaut Training Center. Maezawa, an entrepreneur who made his fortune with a Japanese online apparel retailer, became known in 2018 for his decision to buy a SpaceX Starship circumlunar flight in 2023. He says he still plans to fly on that mission along with eight artists.[10] The Soyuz MS-20 mission to the ISS launched on 8 December from the Baikonur cosmodrome in Kazakhstan, the agency said in a statement, and returned safely on 19 December.

The Soyuz MS-20 mission launched right on schedule and linked up with the ISS six hours (four orbits) after lift-off, docking at the Poisk (Explore) module. After spending almost twelve days aboard the ISS, filming and posting YouTube videos about their experience, Maezawa and Hirano boarded their MS-20 craft on 20 December along with commander Misurkin for the return to Earth. Three hours after a successful undocking they landed southwest of the town of Zhezkazgan in central Kazakhstan, ending Russia's first mission to the ISS carrying two self-funded, non-Russian spaceflight participants in the same Soyuz spacecraft.

Beyond this flight, Roscosmos has announced its flight plan through to mid-2023. The long-term Expedition 67 crew in early 2022 has been confirmed as cosmonauts Oleg Artemyev, Denis Matveyev and Sergey Korsakov. On the following mission, scheduled between autumn 2022 and spring 2023, their backup crew of Sergei Prokopiev, Dmitry Petelin and NASA astronaut Francisco Rubio to fly to the ISS as the Expedition 68 crew. In the latter part of 2023, as part of the ISS-69 expedition, cosmonauts Oleg Kononenko, Nikolai Chub and Andrey Fedyayev have been appointed as the prime crew of the Russian segment of the ISS.[11]

MEANWHILE, ON 21 JULY 2021, the 13-metre (43 ft) *Nauka* (Science) laboratory was launched from the Baikonur Cosmodrome for an eventual link-up with the ISS. Developed by RSC Energia in cooperation with Roscosmos' Khruchinov Centre, it would greatly expand the functionality of the Russian segment of the space station. Docking took place eight days later at an open station port. A short-term problem then occurred due to a software failure which mistakenly initiated a command to turn on the module's jet thrusters for withdrawal, affecting the orientation of the entire orbital complex, causing it to pitch end-over-end at around half a degree per second, or by extension four rotations per hour.[12] As NASA later reported, the mission flight director declared an emergency and his team were finally able to regain the massive station's former stability 45 minutes later by activating thrusters on another module and on a Progress cargo ship previously docked to the ISS. Adding to everyone's concerns, communication with the crew was also intermittent in that time.

Once everything had been stabilized, attention reverted to the *Nauka* module. A reliable internal power and command interface was created, as well as a power supply interface that connected *Nauka* to the space station. Later that day the resident Soviet crew opened the hatches, turned on atmospheric purifiers and began work within the module. There had been some scary moments for the seven crew-members on the station, as well as everyone on the ground, but everything soon settled down and life resumed aboard the ISS.[13]

LIVING ON A MASSIVE, ageing space station is not without its problems and anxieties. The current size and complexity of the ISS is astonishing, comprising of sixteen modules joined in a cross configuration. In late

2021 six modules made up the Russian Orbital Segment, while the American Orbital Segment is made up of eleven modules operated by the United States, the European Space Agency and Japan. As far back as August 2020 a small air leak had been detected aboard the ISS, but tracing the source of the leak remained elusive. At the time, Expedition 63 commander Chris Cassidy reported that

> Both Moscow and Houston Mission Control Centers have been tracking a tiny air leak for several months. A few weeks ago our crew isolated in the Russian segment of space station and closed as many hatches as possible in order to identify the location of the leak. Strangely the data did not point us to any particular location. Yesterday and today, Anatoly [Ivanishin] and I have been checking all of the window seals for any indication of a leak using an ultrasonic leak detector. So far no luck finding the source, but it looks like we will try again with the module isolation this weekend. No harm or risk to us as the crew, but it is important to find the leak [so] we are not wasting valuable air.[14]

On 29 August Roscosmos issued an update on the problem, saying that

> After a thorough analysis and search of the air leak at the International Space Station, the leak was located in the *Zvezda* Service Module containing scientific equipment. The leak is localized in the working compartment of the service module. Currently the search is underway to precisely locate the leak. With that, the general atmosphere pressure decrease rates remain at 1 mm per 8 hours. The situation poses no danger to the crew's life and health and doesn't hinder the station's continued crewed operation.[15]

It would be another six months before repairs could be made on the tiny 22-millimetre (0.86 in.) crack, by which time a second crack had been located. In early March cosmonaut Ryzhikov applied three layers of sealant and placed a patch over the first crack in the Zvezda module. A few days later he applied another layer of sealant, completely covering the patch, and then the second minuscule crack

Marking the fortieth anniversary of the historic Vostok flight of Yuri Gagarin, two Russian cosmonauts bearing the same first name, Yuri Lonchakov and Yuri Usachev, display a photograph of the world's first spaceman in the ISS Service Module, April 2001.

received the same treatment. The repair work was completed on 11 March 2021.

Then, that August, the resident cosmonauts discovered new cracks in the station's Functional Cargo Block (FGB) that could possibly enlarge over time. As Energia's chief engineer Vladimir Solovyov told the RIA news agency at the time: 'Superficial fissures have been found in some places on the *Zarya* module. This is bad and suggests that the fissures will begin to spread over time.' Solovyov did not disclose whether any air had leaked out. He had previously expressed his opinion that much of the station's equipment was ageing and warned there could be an 'avalanche' of broken equipment after 2025.[16]

Despite these reservations, Roscosmos confirmed that it will remain part of the ISS programme until 2024 and is tentatively amenable to extending its participation beyond that point. However, a decision will then be made on any future participation in the ISS as its technical modules will be approaching the end of their service life, and there are plans and enthusiasm mounting for Russia to develop its own orbital outpost, although it would be a monumentally expensive undertaking. Deputy Prime Minister Yuri Borisov is unconvinced that Russia should continue to ferry people to the aged ISS. 'We can't risk the lives [of our cosmonauts],' he stated on Rossiya 1 state television. 'The situation that today is connected to the structure and the metal getting old, it can lead to irreversible consequences – to catastrophe. We mustn't let that happen.'[17] There are certainly many crucial decisions to be made and implemented before 2024 rolls around.

ON 12 APRIL 2021 Russia proudly marked six decades of human space flight in celebrating the anniversary of Yuri Gagarin's Vostok mission, during which he became the first person to ever fly into space and into orbit on that truly historic morning of 12 April 1961.

REFERENCES

PROLOGUE

1 Alexander Zheleznyakov, 'Konstantin Ivanovich Konstantinov', *Orbit, Journal of the Astro Stamp Society* (October 2004).
2 'Tsiolkovsky, Konstantin Eduardovich (1857–1935)', http://weebau.com, accessed 14 August 2020.
3 'Hermann Oberth', www.nasa.gov, 22 September 2010.
4 'Konstantin Eduardovich Tsiolkovsky', www.encyclopedia.com, accessed 14 August 2020.
5 Colin Burgess and Chris Dubbs, *Animals in Space: From Research Rockets to the Space Shuttle* (Chichester, 2007), pp. 12–22.
6 Konstantin Tsiolkovsky quoted in Francis French and Colin Burgess, *Into that Silent Sea: Trailblazers of the Space Era, 1961–1965* (Lincoln, NE, 2007), p. 18.

1 PUPNIKS AND SPUTNIKS

1 Colin Burgess and Chris Dubbs, *Animals in Space: From Research Rockets to the Space Shuttle* (Chichester, 2007), p. 61.
2 Ibid., p. 63.
3 Ibid., pp. 66–8.
4 Alice E. M. Underwood, 'The First Canine Cosmonauts', www.russianlife.com, 22 July 2016.
5 A. V. Podrovsky, 'Vital Activity of Animals during Rocket Flights into the Upper Atmosphere', in *Behind the Sputniks: A Survey of Soviet Space Science*, ed. F. J. Krieger (Washington, DC, 1958). Originally presented as a report to the International Congress on Guided Missiles and Rockets, Paris, 3–8 December 1956.
6 John Rhea, ed., *Roads to Space: An Oral History of the Soviet Space Programme* (London, 1995), p. 295. Compiled by the Russian Scientific Research Centre for Space Documentation. Translated by Peter Berlin, Aviation Week Group.
7 Dmitry C. Malashenkov, 'Some Unknown Pages of the Living Organism's First Orbital Flight', Institute for Biomedical Problems (Moscow). Eight-page paper presented at the 53th International Astronautical Congress. The World Space Congress, Houston, Texas – 2002, 10–19 October 2002.
8 Colin Burgess and Chris Dubbs, *Animals in Space: From Research Rockets to the Space Shuttle* (Chichester, 2007), pp. 204–7.
9 Ibid., pp. 212–13.

2 'POYEKHALI!'

1 Colin Burgess and Rex Hall, *The First Soviet Cosmonaut Team: Their Lives, Legacy, and Historical Impact* (Chichester, 2009), pp. 42–3.
2 Ibid., p. 44.
3 Ibid., p. 45.
4 Anna Smolchenko, 'Russia Remembers Horrific Space Accident', *Sydney Morning Herald* (25 October 2010), p. 14.
5 'Obituary: Space-Race Pioneer Oleg Ivanovsky', www.smh.com.au, 26 September 2014.
6 Francis French and Colin Burgess, *Into That Silent Sea: Trailblazers of the Space Era, 1961–1965* (Lincoln, NE, 2007), p. 19.
7 'The Flight of Vostok 1', www.esa.int, accessed 24 November 2020.
8 Robin McKie, 'Sergei Korolev: The Rocket Genius behind Yuri Gagarin', www.theguardian.com, 13 March 2011.
9 Paul Rodgers, 'Yuri Gagarin: The Man Who Fell to Earth', www.independent.co.uk, 23 October 2011.

3 VOSTOK FLIGHTS CONTINUE

1 Francis French and Colin Burgess, *Into that Silent Sea: Trailblazers of the Space Era, 1961–1965* (Lincoln, NE, 2007), p. 105.
2 Colin Burgess and Rex Hall, *The First Soviet Cosmonaut Team: Their Lives, Legacy and Historical Impact* (Chichester, 2009), p. 174.
3 Anatoly Zac, 'Vostok-2 Mission: First Day in Space', www.russianspaceweb.com, accessed 24 November 2020.
4 Lester A. Sobel, ed., *Space: From Sputnik to Gemini* (New York, 1965), pp. 123–4.
5 Ibid., p. 126; selected quotes from Dr Vladimir Yazdovsky, Moscow's Institute of Aviation Medicine (IAM), and Dr Oleg Gazenko, Soviet Academy of Sciences, at 12th Annual Congress of the International Astronautical Federation, Washington, DC, 4 October 1961.
6 Michael Klesius, 'Sick in Space', www.airspacemag.com, 8 March 2009;
7 'Spaceman's Secret Out: Landed by Parachute', *Daily Mirror* (Sydney) (4 May 1962), p. 6.
8 'Zenit-2 Satellite: Part of Vostok', www.astronautix.com, accessed 22 December 2020.
9 Umberto Cavallaro, *The Race to the Moon: Chronicled in Stamps, Postcards, and Postmarks* (Chichester, 2018), pp. 138–9.
10 Burgess and Hall, *The First Soviet Cosmonaut Team*, p. 195.
11 Bart Hendrickx, 'The Kamanin Diaries, 1960–1963', *Journal of the British Interplanetary Society*, 50 (London, 1997), pp. 33–40.
12 Burgess and Hall, *The First Soviet Cosmonaut Team*, pp. 203–7.
13 Ibid., p. 231.
14 Reina Pennington, 'Tereshkova, Valentina (1937–)', www.encyclopedia.com, accessed 14 December 2020.
15 A. Lothian, *Valentina, First Woman in Space: Conversations with A. Lothian* (Durham, 1993), p. 231.
16 Hendrickx, 'The Kamanin Diaries, 1960–1963', pp. 33–40.

4 SOVIET SPECTACULARS AND A SPACEWALK

1 Mark Wade, 'Voskhod 1: Part of Vostok', www.astronautix.com, accessed 26 November 2020.
2 Asif Siddiqi, *Challenge to Apollo: The Soviet Union and the Space Race, 1945–1974* (Washington, DC, 2000), pp. 421–2.
3 Francis French and Colin Burgess, *Into that Silent Sea: Trailblazers of the Space Era, 1961–1965* (Lincoln, NE, 2007), pp. 344–7.
4 David Szondy, 'A Step Back in Time: The 50th Anniversary of the First Spacewalk', https://newatlas.com, 19 March 2015.
5 Siddiqi, *Challenge to Apollo*, pp. 452–3.
6 Alexei Leonov, interview with Fédération Aéronautique Internationale (FAI), on fiftieth anniversary of EVA, www.collectspace.com, Sourced under collectSPACE Space News Archive, 18 March 2015 at www.collectspace.com.
7 Ajai Raj, 'The Terrifying Story of the First Person to Walk in Space – and How He Almost Didn't Make It Back', www.businessinsider.com.au, 18 October 2014.
8 Paul Rincon, 'The First Spacewalk', www.bbc.co.uk, 13 October 2014.
9 Ibid.
10 Ibid.
11 Wade, 'Voskhod 1: Part of Vostok'.

5 THE TROUBLE WITH SOYUZ

1 Colin Burgess and Kate Doolan, *Fallen Astronauts: Heroes who Died Reaching for the Moon* (Lincoln, NE, 2016), pp. 244–5.
2 Eugen Reichl, *The Soviet Space Program: The Lunar Mission Years, 1959–1976* (Stuttgart, 2019), p. 100. Translated from German by David Johnston.
3 Asif A. Siddiqi, *Challenge to Apollo: The Soviet Union and the Space Race, 1945–1974* (Washington, DC, 2000), p. 587.
4 Burgess and Doolan, *Fallen Astronauts*, p. 245.
5 Brian Harvey, *Soviet and Russian Lunar Exploration* (Chichester, 2007), p. 88.

6 LOSING THE MOON

1 Francis French and Colin Burgess, *Into that Silent Sea: Trailblazers of the Space Era, 1961–1965* (Lincoln, NE, 2007), pp. 33–4.
2 Mark Wade, 'Beregovoi, Georgi Timofeyevich', www.astronautix.com, accessed 26 December 2020.
3 Rex Hall and David Shayler, *Soyuz: A Universal Spacecraft* (Chichester, 2003), p. 145, quoting from 'Potholes on the Starry Road: An Interview with Cosmonaut Konstantin Feoktistov', *Trud* (April 2002).
4 David Whitehouse, 'How the USSR Nearly Beat the Americans to the Moon', *Canberra Times* (14 April 1986), p. 2.
5 James Oberg, 'The Moon Race Cover-Up', *Reason* (August 1979), at https://reason.com.

6 David Scott and Alexei Leonov, *Two Sides of the Moon* (London, 2004), pp. 187–8.
7 Mark Wade, 'Soyuz 7K-L1: Part of Soyuz', www.astronautix.com, accessed 26 December 2020.
8 Nikola Krastev, 'Soviet Cosmonauts Recall Failed Bid to Beat U.S. to Moon', www.rferl.org, 21 July 2009.
9 'Moon Landings: Soviet Abandons Plans', *Canberra Times* (27 October 1969), p. 6.
10 Scott and Leonov, *Two Sides of the Moon*, p. 254.

7 A TRAGIC SETBACK

1 David M. Harland, 'Salyut', www.britannica.com, accessed 28 December 2020.
2 Eric Thornton, 'Soviet Feat Lift to Soviet Science', *Tribune* (February 1969), p. 12.
3 David Scott and Alexei Leonov, *Two Sides of the Moon* (London, 2004), p. 261.
4 Bernard Gwetzman, '3 Soviet Astronauts are Dead; Bodies Discovered in Capsule When it Lands After 24 Days', *New York Times* (June 1971), p. 1.
5 Ibid., p. 1.

8 DÉTENTE IN ORBIT

1 'Kennedy Proposes Joint Mission to the Moon', www.history.com, 13 November 2019.
2 President John F. Kennedy, speech delivered in person before a joint session of Congress, 25 May 1961, www.jfklibrary.org.
3 'The Apollo-Soyuz Mission', www.nasa.gov, 18 March 2010.
4 Francis French, 'Imagining a World Where Soviets and Americans Joined Hands on the Moon', www.smithsonianmag.com, 19 July 2019.
5 Matthew Bodner, 'Russian and U.S. Space Legends Meet 40 Years after "Handshake in Space" (Video)', www.themoscowtimes.com, 16 July 2015.

9 SPACE STATION MIR

1 Mark Wade, 'Salyut 4: Part of Almaz', www.astronautix.com, accessed 5 January 2021.
2 Mark Wade, 'Salyut 5: Expeditions 1 and 2' and 'Almaz OPS: Part of Almaz', www.astronautix.com, accessed 31 July 2021.
3 Doug Adler, 'The Forgotten Rescue of the Salyut 7 Space Station', https://astronomy.com, 23 October 2020.
4 John T. McQuiston, 'Salyut 7, Soviet Station in Space, Falls to Earth After 9-Year Orbit', *New York Times* (7 February 1991), section A, p. 8.
5 Harvey Day, '"Three-foot flames and melting metal": How I Survived the Biggest Ever Fire in Space' (interview with Jerry Lineger), www.shortlist.com, 3 May 2018.
6 Marcia Dunn, 'Dousing Fire on Mir a Matter of Survival', *Lakeland Ledger* (21 March 1997), p. A2.

7 Kathy Sawyer, 'Docking Crash Cripples Mir Space Station', www.washingtonpost.com, 26 June 1997.

10   THE SOYUZ LEGACY

1 Robert Pearlman, 'STS-135: The Final Flight', www.collectspace.com, accessed 30 November 2020.
2 'Russia Declares "Era of Soyuz" after Shuttle' (uncredited article), https://phys.org, 21 July 2011.
3 Ibid.
4 Richard Hollingham, 'Soyuz: The Soviet Space Survivor', www.bbc.com, 2 December 2014.
5 Ibid.
6 Ibid.
7 Jeff Foust, 'Soyuz Marks Ends of an Era for NASA', www.spacenews.com, 14 October 2020.
8 Tony Quine, *Russian Movie in Space, Part 2*, Space Sleuthing blog, 'Russian Actresses Who Will Compete for Trip to ISS', www.spacesleuth2.blogspot.com, 21 March 2021.
9 Tony Quine, *Russian Movie in Space, Part 10*, Space Sleuthing blog, 'Russian "Movie in Space" Preparations Enter the Final Month', www.spacesleuth10.blogspot.com, accessed 2 September 2021.
10 Soviet–Russian Space forum at www.collectspace.com: Space Adventures release: 'Space Adventures Client, Yusaku Maezawa, Plans for Mission to the International Space Station', www.collectspace.com, accessed 13 May 2021.
11 Russian Aviation, 'Roscosmos Appointed ISS Crews until 2023', www.ruaviation.com, 20 May 2021.
12 Roscosmos news release, 'Energia Designer General on MLM Nauka Docking', 30 July 2021, accessed from collectSPACE Messages forum at www.collectspace.com.
13 Steve Gorman and Polina Ivanova, 'International Space Station Thrown Out of Control by Misfire of Russian Module – NASA', *Reuters Science*, www.reuters.com, 30 July 2021.
14 Space Shuttles –Space Station forum at www.collectSPACE.com: 'From Expedition 63 Commander Chris Cassidy', www.collectSpace.com, 24 September 2021.
15 Roscosmos news release, 'Energia Designer General on MLM Nauka Docking'.
16 Gabrielle Tétrault-Farber, 'Russian Cosmonauts Find New Cracks in ISS Module', *Reuters Science*, www.reuters.com, 30 August 2021.
17 BBC News, uncredited article, 'International Space Station Facing Irreparable Failures, Russia Warns', www.bbc.com/news, 1 September 2021.

FURTHER READING

Baker, David, *The History of Manned Space Flight* (New York, 1981)
Burgess, Colin, and Kate Doolan, *Fallen Astronauts: Heroes who Died Reaching
    for the Moon* (Lincoln, NE, 2003)
——, and Chris Dubbs, *Animals in Space: From Research Rockets
    to the Space Shuttle* (Chichester, 2007)
——, and Rex Hall, *The First Soviet Cosmonaut Team: Their Lives, Legacy and
    Historical Impact* (Chichester, 2009)
Doran, Jamie, and Piers Bizony, *Starman: The Truth Behind the Legend of Yuri
    Gagarin* (London, 1998)
French, Francis, and Colin Burgess, *Into that Silent Sea: Trailblazers of the
    Space Era, 1961–1965* (Lincoln, NE, 2007)
——, *In the Shadow of the Moon: A Challenging Journey to Tranquility*
    (Lincoln, NE, 2007)
Gagarin, Yuri, *Road to the Stars* (Moscow, 2002)
Hall, Rex, and David J. Shayler, *Soyuz: A Universal Spacecraft*
    (New York, 2003)
——, —— and Bert Vis, *Russia's Cosmonauts: Inside the Yuri Gagarin Training
    Center* (New York, 2007)
Harvey, Brian, *Russia in Space* (New York, 2001)
——, *The Rebirth of the Russian Space Program: 50 Years after Sputnik, New
    Frontiers* (New York, 2007)
Jenks, Andrew L., *The Cosmonaut who Couldn't Stop Smiling: The Life and
    Legend of Yuri Gagarin* (DeKalb, IL, 2012)
Siddiqi, Asif, *Challenge to Apollo: The Soviet Union and the Space Race,
    1945–1974*, NASA History Series SP-2000-4408 (Washington, DC, 2011)
Tsymbal, Nikolai, ed., *First Man in Space: The Life and Achievements of Yuri
    Gagarin* (Moscow, 1984)
Walker, Stephen, *Beyond: The Astonishing Story of the First Human to Leave
    Our Planet and Journey into Space* (London, 2021)